## STAR WARS:
## THE NEW HIGH GROUND

This book describes the job the US military is already doing in space, takes a look at military space technology now on the horizon, and evaluates some of the proposals for turning space into the combat arena of the future.

When you press the most ardent spacemen – and some of their civilian sympathizers – for the details on just how we might obtain a decisive military advantage from space, they often point to the potential of laser weapons. Can a space laser battle station, at a cost of some hundreds of billions of dollars, free us from the nuclear balance of terror by shielding us from Soviet weapons? It's not likely.

It is the dual thesis of this book that although it is indeed time that we recognize the military usefulness of space technology, we need also to understand that space power is not going to provide us with a military superiority that will solve all our problems – any more than air power did before it.

## About the Author

Thomas Karas received his Ph.D. from Harvard University in 1971. He has been a Consulting Research Associate for the Federation of American Scientists and a Consultant to the US Congress.

As Professor of International Relations at Boston University and a senior analyst for defence information, Dr Karas is excellently qualified to compile this first major report on space-war weapons and strategies, for which he interviewed officials from the Pentagon, the Aerospace Defense Command, and the AF Defense Division in Los Angeles.

# Star Wars: The New High Ground

## Systems and Weapons of Space Age War

---

# Thomas Karas

**NEW ENGLISH LIBRARY**
Hodder and Stoughton

First published in the United States of America in 1983 by Simon and Schuster

First published in Great Britain in 1984 by New English Library

First New English Library Paperback edition 1988

**British Library C.I.P.**

Karas, Thomas
 [The new high-ground] Star wars: the new high ground: systems and weapons of space age war
 1. Space warfare
 I. [The new high-ground]
 II. Title
 358'.8

ISBN 0-450-42401-4

Printed and bound in Great Britain for Hodder and Stoughton Paperbacks, a division of Hodder and Stoughton Ltd., Mill Road, Dunton Green, Sevenoaks, Kent TN13 2YA (Editorial Office: 47 Bedford Square, London WC1B 3DP) by Richard Clay Ltd., Bungay, Suffolk.

To George and Corrine Karas

# Contents

# The Spacemen

Two hundred and fifty spacemen held a convention in the spring of 1981 at the Air Force Academy. A year earlier Dr Hans Mark, then Secretary of the Air Force, had suggested that it was time the Air Force had an official 'space doctrine' as well as 'air doctrine.' A group of young officer-professors in the Academy's Department of Astronautics and Computer Science quickly formed a 'Space Doctrine Group' at the Academy. They organized a course on space doctrine, and sent out a call for the first 'Military Space Doctrine Symposium.'

Military men could hardly have picked a more suitable place than the Air Force Academy to talk about space and the future. The Academy campus, several miles north of Colorado Springs, has a futuristic air. The main, or 'cadet,' area stands on the side of a mountain, one of the Rampart Range. From a distance, it looks like an illustration for a science-fiction paperback. Up close, it looks more like a fortress. Not, like West Point, a crenellated fortress of old, gray, ivy-covered stone, but an edifice of massive interlocking blocks of concrete, steel, aluminium, and glass. Walking up along the ramparts or down between buildings, first-year cadets 'square' the corners, turning at sharp right angles as though guided by an invisible force field.

The organizer of the meeting, Major Charles Friedenstein, who worked in the Academy's Department of Astronautics and Computer Science, is a slim, intense, scholarly-looking young man. He explained the purpose: 'We were trying to see if we could get the choir singing the same tune.' The 'choir' in this case is a coterie of junior and middle-level (lieutenant to colonel) Air Force officers who believe strongly in the future of 'space power.' The congregation is the rest of the Air Force.

The organizers also did what they could to keep the symposium a closed choir practice rather than a public performance. Aside from some big brass in the Air Force and the spacemen, they invited one or two academic experts, a few analysts from Air Force space contractors, and a couple of Navy spacemen. The conference was not classified 'confidential' or 'secret,' but observers from outside the military space community were not welcome. 'We wanted people to be able to speak their minds freely,' said Friedenstein.

Major Friedenstein's 'choir' metaphor exemplifies the spacemen's almost religious dedication to their cause. Major Peter Swann, the editor of the symposium background papers, regretted that there were no copies available of the symposium readings, but suggested I read a book: *A Few Great Captains* by DeWitt S. Copp, published by Doubleday in 1980. 'We will have to learn from history,' said Swann, 'or the country will be on its knees again the way it was before World War Two.'

A few weeks later I visited another officer at the Pentagon to discuss a paper he had prepared for the symposium and he began by recommending *A Few Great Captains*. The next officer I interviewed walked into the room carrying the book. Another displayed it prominently on a lamp table in his office. This book began to look like the spacemen's bible. I bought it at the Pentagon book store.

*A Few Great Captains: The Men and Events That Shaped the Development of U.S. Air Power* is dedicated 'To all the U.S. Army airmen, few in number, great in spirit – the seekers, the pathfinders, the builders. They dared the heights and saw beyond their time.' It is a history (to September 1939) of the careers of five Army Air Corps officers who became the patriarchs of the U.S. Air Force: Frank Andrews, Hap Arnold, Ira Eaker, Benny Foulois, and Tooey Spaatz. It is an epic tale of their long struggle to separate the Air Force from the Army.

The prominent prophet of *A Few Great Captains* is Billy Mitchell. Mitchell's prophecy (and the dogma for which the patriarchs struggled) was that air power was the weapon of the future. The book describes Mitchell's role in the years between the end of World War I and his court-martial in 1925: 'There was an electricity to the man that inspired and lit up the thinking of his airmen and sent sparks flying among those who opposed him. He had become a convinced champion of and the leading U.S. military spokesman for an air force independent of General Staff control. To this belief Arnold, Spaatz and most of their fellow airmen had become firmly dedicated.'

In 1921 Mitchell made headlines for air power by organizing the bombing and sinking of a former German battleship, the *Ostfriesland*. As assistant to the Army Air Service Chief in 1923 he ran another such demonstration. At the National Aeronautic Convention in 1924, Mitchell '. . . spoke on the unmentionable subject of strategic bombardment and its effect on an adversary's means of production.' This was the first time he had raised the theory publicly, and he was the first American airman to do so. He had already proclaimed that navies were going out of business. Now he suggested, '. . . that aerial bombardment properly planned might make it unnecessary for armies to meet on the battlefield.'

Mitchell's unrelenting publicity campaign for air power and resistance to the powers-that-were finally got him court-martialed and driven from the Army in 1925. Among other offenses, he charged the Navy and War Departments with '. . . incompetency, criminal negligence, and almost treasonable administration of the National Defense . . .' The Army tribunal found Mitchell guilty of 'conduct of a nature to bring discredit upon the military service.' The prophet of air power had sacrificed himself for his beliefs – fallen on his sword. It was left to the patriarchs, the young airmen who stayed inside the Army, to struggle on.

If *A Few Great Captains* is about the rise of *air* power, how did it come to be the *space*man's bible? The theory of air power held that the decisive weapon of the future would be the bomber, which would destroy the enemy's ability to make war by ruining the industrial plant and lines of transport which fed armies and navies. As we'll see later, while the spacemen believe in the great military potential of space, and some even believe that dominant military power over the earth can be achieved from space, the spacemen do not all share a clear-cut, highly developed doctrine in the way that the early airmen did. What draws the spacemen to *A Few Great Captains* is that the air-power pioneers finally did succeed in liberating the Air Force from the ground and sea services.

Major Roger Dekok, an Air Force Space Systems Staff officer at the Pentagon, said, 'You don't have to be a military historian to look at the fact that when new environments have been opened up, that contrary to the wishes of well-meaning people, they've always been exploited for military purposes. We don't believe the United States can afford to be second in that sort of exploitation. The people who were looking at air power in the twenties and thirties [were] facing the same sort of difficulty in getting the leaders to realize the potential of air power to change the course of war. And it took World War Two to change that and we don't want World War Three to be the end of this.'

One of Dekok's colleagues in the Pentagon space operations office, Major William Savage, said, 'The space guys find hope in this book – see guys with the same problems winning out in the end. The Air Force is not acting terribly unlike the Army did some thirty or forty years ago when the Air Corps people were trying to get their airplanes developed and the Army people, who were more familiar with artillery, tanks, and – back in those days – horse cavalry, they and people on up to the President were saying, "Ah, get one airplane and let those flyboys take turns flying it." To a degree, the space guys have the same problem: most of the people are more familiar with airplanes than they are with space. Every penny that goes into Air Force Space comes out of the Air Force budget, it comes out of some other Air Force program. It's hard to get the money to do all the things that the space guys think are necessary.'

Although the 'space guys' are not unanimous on all points, they do agree on several ideas. They think the time has come to treat space more as an arena of military operations and not merely as an area for research and development. They say we need to plan for and spend more money on the weapons for space that are now on the technological horizon. Looking beyond, they want a more visionary, daring approach to future possibilities for the military exploitation of space. Many of them believe that the United States can use space to obtain a substantial, if not decisive, military advantage over the Soviet Union (and that if we don't get that advantage, the Soviets will).

Many see the immediate obstacle as organizational: just as the Army Air Corps was stifled by the old Army ways of thinking and acting, so those working on space are hindered by the narrow, if understandable, focus of the Air Force on air power. Some of the spacemen proposed that there should be an operational 'specified command' for space within the Air Force, a command that would not be treated merely as a support or service organization for the 'real' combatants, but as a coequal unit in command of its own 'forces.' It would have an equal voice with other operational commands in the councils of the Air Force Chief of Staff office and it would have its own, separate budget. In September 1982, the Air Force did create a Space Command that at least began to do what the spacemen wanted.

Other spacemen see an eventual need for a U.S. Space Force. They believe space is as different a medium for military operations from the air as the air from the sea or the sea from the land. Major Savage does not necessarily advocate a separate space force, but he does point out the analogy many see in *A Few Great Captains*: 'The eventual organizational evolution in the case of the Army Air Corps versus the Army was a separate service. This may not have been necessary and it may not be necessary that a separate service be established in the space area, although parallel logic would have you think so, since the services are based on the media in which operations are conducted . . . but I think that if the Air Force got its act together, put its heart and its funds into it, that we can do a heck of a good job in space.'

This book describes the job the U.S. military is already doing in space, takes a look at military space technology now on the horizon, and evaluates some of the proposals for turning space into the combat arena of the future.

Chapter One, 'The Space Establishment,' is an introduction to how the Pentagon, and particularly the Air Force, is organized to operate in space. Recent, as well as proposed, changes there should encourage the spacemen. There is growing recognition in the Pentagon of the military importance of space, and the Air Force does indeed seem to be tooling up to move combat beyond the earth's atmosphere. One stimulus for the growing military attention to space has been the space shuttle. The prospect it offers of regular transportation of heavy loads into earth's orbit has stirred the imaginations of military space planners.

Chapter Two, 'The Space Business,' looks at the space shuttle, or National Space Transportation System, and introduces the industry that supplies the stuff of which military planners' dreams are made. It's an industry of paradox: capable of producing prodigious technological marvels, it seems incapable of meeting contractual cost and time limits; endowed with thinkers even more visionary and excited by opportunities in space than the military spacemen, it seems all too ready to grab an extra buck at the taxpayer's expense.

Chapter Three, Four and Five describe the hardware the space industry provides. More than most people – in or outside the military – recognize, our military forces have become dependent on satellites to support their work. In communications, reconnaissance, weather forecasting, and geodetics and navigation, space systems let military men do what would otherwise be technologically and economically impossible. It's easy to see why the spacemen feel unjustly neglected: the support they provide to their comrades-in-arms is pervasive, yet taken for granted.

Chapter Six, 'ASAT: Weapons in Space,' and Chapter Seven, 'War in Space?' point out that with the advantages offered by the growing military exploitation of space come new risks as well. Support from space means dependence *on* space. The Soviet Union has been working for years – with some success – on weapons to knock out U.S. satellites. The more dependent our military forces become on satellites, the more tempting those spacecraft will be as military targets. The Air Force, for its part, is interested not only in defending satellites, but in being able to attack Soviet ones: the new U.S. anti-satellite weapon, to be space-tested in 1983, should be far superior to anything the Soviets have shown yet.

Chapter Eight shows that it remains to be seen whether military planners can come up with weapons or defensive measures that could adequately protect space assets from enemy attack. The difficulty of defending satellites might suggest that arms control measures negotiated with the Soviets might be an alternative means of keeping the space lanes open. But most U.S. military men who think about this question are skeptical about arms control in space. Not only do they doubt that the verification of agreements could be 100 percent foolproof, but they are intrigued with the possibility of the U.S. attacking valuable *Soviet* military satellites in wartime. We are, in short, well on the way to extending into space the expensive arms race we've been running on the earth.

When you press the most ardent spacemen – and some of their civilian sympathizers – for the details on just how we might obtain a decisive military advantage from space, they often point to the potential of laser weapons, which we will also explore in the final chapter. Can a space laser battle station, at a cost of some hundreds of billions of dollars, free us from the nuclear balance of terror by shielding us from Soviet weapons? It's not likely.

It is the dual thesis of this book that although it is indeed time that we recognize the military usefulness of space technology, we need also to understand that space power is not going to provide us with a military superiority that will solve all our problems – any more than air power did before it.

# CHAPTER ONE

# The Space Establishment

For the spacemen who had gathered at the Air Force Academy in the spring of 1981, the year following their symposium must have been heartening: events began to move in their direction. Defense officials in the new Reagan Administration proved sympathetic to their cause. The Air Force began to give organizational recognition to the military importance of space. Support for a Space Command emerged from Congress. Military space budgets were up.

In September 1981, Air Force headquarters at the Pentagon opened up a new space shop. The Deputy Chief of Staff for Plans and Operations and Readiness added a fifth subunit to its organization chart – the Directorate for Space Operations. The Deputy Chief of Staff, Lieutenant General Jerome O'Malley, had himself been sympathetic to the spacemen's view of the significance of military space programs. Now, for the first time, he had under his command a brigadier general and a thirty-man staff specifically in charge of space.*

---

* A year earlier, O'Malley had hinted at the direction his office would take by telling the group assigned to the Houston Space Center to work on the shuttle: 'We must apply the same considerations to space systems as we do for other operations. We must design space assets, and structure their supporting organization in a manner responsive to the needs of operational forces – and integrate them into these forces – to allow field commanders to be confident that space capabilities will be there when they are needed.'

In that same speech, O'Malley said the other things the spacemen love to hear: 'I believe the use of space by military forces is at a point paralleling the position of air power after WW I.' And: 'While peace may be our profession, being prepared for war is our business, and we must be prepared to protect our vital interests in space as well as those in land, sea, and air.' And: 'The potential for space to become a more hostile environment is increasing. It is increasing for the very reasons that air became an arena for hostilities: first, because space systems provide increasingly important support – some would say a decisive edge – to military forces; and second, the technology for space conflict is available.' The most dedicated spacemen couldn't have said it better himself.

With the new directorate the spacemen had a platform for spreading the space gospel through the Air Force and the Pentagon at large. The directorate, headed by Brigadier General John H. Storrie (a reconnaissance and electronic warfare expert), would begin drawing up broad Air Force plans for space.

It would push the Air Force to take full advantage of the potential of the space shuttle. It would brief other Pentagon officers on the threat of the Soviet space program and the promise of U.S. military space activities. It would try to influence the Reagan Administration's revisions of national space policy. And it might even become the first step toward an Air Force Space Command.

The conferees at the Academy space symposium had been divided over whether the time was ripe for a Space Command. Trying to push it out too soon could turn the rest of the Air Force bureaucracy against the space cause. But in 1981 and 1982 outside pressures for the Space Command idea began to grow. In November 1981, the Under Secretary of the Air Force, Edward Aldridge, Jr., told the National Space Club that the operation of U.S. military space assets had to be better coordinated in the future: 'I believe the right answer may be some form of a "space command" for the *operation* of our satellites and launch services.' He added, 'The Air Force is moving in that direction now.'

One step was the creation of a 'Defense Space Operations Committee, chaired by the Secretary of the Air Force, to coordinate space operations activities for DoD.' Another was the opening of the Air Force Directorate for Space Operations. Another was the creation of a course in 'space operations' at the Air Force Institute of Technology. Yet another was the appointment of an Air Force general, James Abrahamson, as Associate Administrator of NASA for Space Transportation Systems, responsible for the shuttle. While NASA represented this appointment as merely filling the job with the best available manager, the Air Force Under Secretary boasted that Abrahamson '. . . still wears his "Blue" suit.'

Over on Capitol Hill, some members of Congress also

pushed the Air Force to move toward a Space Command. Representative Ken Kramer of Colorado Springs, a full convert to the spacemen's point of view, introduced a bill late in 1981 to rename the Air Force the 'United States Aerospace Force.' The bill, House Resolution 5130, would require the Aerospace Force '. . . to be trained and equipped for prompt and sustained offensive and defensive operations in air and space, including coordination with ground and naval forces and the preservation of free access to space for U.S. spacecraft.' It would also direct the '. . . Secretary of the Aerospace Force to report to Congress on the feasibility of establishing a separate space command." Official Air Force comment on the bill was unfavorable.

Early in 1982 Space Command supporters got more help from the Congressional 'watchdog,' the General Accounting Office. The GAO advised Congress to limit funds for the Air Force's planned Consolidated Space Operations Center (CSOC) until the Defense Department came up with an overall plan for the military exploitation of space. The report said that the CSOC 'could be used as the nucleus for a future space force' or for a 'future space command,' and that it ought to be planned with that in mind.

Defense Department civilian officials responded that space was a 'place,' not a 'mission,' that 'space systems compete with other types of systems in establishing the most effective means of accomplishing a given mission.' But for the spacemen, the GAO report was extra ammunition.

Then in June 1982, the Air Force announced that it was creating a Space Command, to start up the following September. The Space Command would begin with jurisdiction over only some space activities, but it was expected to form the core of an increasingly unified military space program.

Even without the platform of a Space Command from which to campaign for bigger slices of the budget pie, the spacemen had made financial progress in 1982. The Department of Defense request for fiscal 1983 military space appropriations (not counting those for super-secret spy satellites) came to $8.5 billion, compared with the $5.8 requested for fiscal 1982. Before inflation, that was an increase of more than 46 percent. In the previous two years, the increases had been 21 and 24 percent. In each year, these increases ran well ahead of the growth of the whole Defense budget.

I spoke with the deputy director of the new Air Force Space Directorate just before he moved into that new position. Colonel Earl Van Inwegen is a blond, athletic-looking man with a powerful handshake and a friendly smile. He was then head of a Space Operations Division that was subordinate to the Directorate of Operations and Readiness. Clearly that work of that division would carry over into the new Space Directorate. 'We're the operations focal point for Air Force space activities,' said the Colonel. 'In some programs we have formal responsibility, in others we're trying to coordinate the developers on the one hand and the users on the other. We're trying to get the space systems closer to what's actually needed by the operators – the commanders and organizations who use the services that the satellites provide. This is a slow process, because satellites are long-lived. It's hard to change them: it's not like bringing the F-15 [jet fighter] fleet in and making a mod on it – you've got to wait until the next block change or the next satellite goes up, and that could be three or four years.'

Originally space systems had been almost exclusively the responsibility of the research and development side of the Air Force bureaucracy. 'In the past year or so we've been getting more formal responsibility here on the operations side of the house,' said Van Inwegen. 'You can liken it to the way the Air Force runs their more "normal" programs. When the F-15 or the F-16 or the B-1 bomber is being developed, the focus is on the research and development side. And then when that system becomes operational, the responsibility for its maintenance and use goes to the logistics guys and the operations people, and the R-and-D guys start working on something else. In the space business that hasn't happened in the past. The R-and-D guys build the system, it becomes operational, and they continue to function as the point of contact in the headquarters for operating and maintaining that system.'

But what undesirable effects does this 'operational' role of the research and development people have? Van Inwegen offered the example of Air Force work on anti-satellite weapons: 'The Soviets are way ahead of us in that area. Not that we haven't been working on it, it's just this technological enamorment we have – we want to make the thing the best, the most sophisticated . . . It's a function of the R-and-D involvement: their job is to climb technical hurdles, make things new and different, whereas the operator just wants something that'll work, do the job. The longer the program carries on, the more complex it's going to get. They're always hanging that technological carrot out in front of you: "If we just did this, it would be so much better and be able to meet your requirements in a much more responsive manner." But you never have it *now* – it's always some day in the future.'

I asked Van Inwegen if he was a fan of *A Few Great Captains*. 'Yes! You see a parallel in there with the R-and-D thing: the Air Corps men were very upset about the Army Materiel Command running the airplane business, and we kind of say the same thing about the research and development community.' He pointed to a framed cartoon on his alcove wall. It portrayed the busts of two figures. The figure on the left was a man in World War I U.S. Army uniform, heavily decorated. Above the figure was written AIRPOWER, and below, the name 'Billy Mitchell.' The figure on the right was in Air Force uniform, but it had a question mark where its face should have been. Above was printed SPACEPOWER, and below another question mark.

Like many of his colleagues, Van Inwegen wanted to see a special Space Command in the Air Force. In the meanwhile, he thought the Air Force could still move ahead more rapidly in the military exploitation of space. 'Let's press on: we have the infrastructure to be able to move forward now. Now there's a lot of political infighting that goes on because of the lack of a focused organization, but it's probably going to be that way even when you get one.

You always get people who want to change organization.'

As he saw it, one of the obstacles to the fuller military exploitation of space is a lack of recognition in the Air Force as a whole for the importance of space. 'There are several reasons for that. One is that very few of our senior leadership have been involved in the space systems. With so many private contractors running the space systems, there are relatively few "blue-suiters," Air Force officers involved except in the management of R-and-D programs. So the kind of people that move up in the space chain are the guys in the research and development business, the engineers, and they're kind of off to the side of the Air Force mainstream of officer promotion.

'And then a lot of things having to do with space are classified, and only those with a "need to know" ever hear anything about them. Those who don't [hear] get defensive: "If you're not going to tell me about it, to hell with you." They turn it off and don't *want* to know about it.

'How many people have patted a satellite on the head? Everybody goes out and flies an F-15, a B-52. I call it the "stick and rudder psychosis." The senior leadership knows what it's like to fly an airplane: they can appreciate "pulling g's", and turning right and left. We say, "You push the stick down, the houses get bigger, you pull it up, the houses get smaller." The leaders just aren't familiar with space systems.

'I think people take space for granted. How many times do people hear, "This is brought to you direct by satellite"? They take it as a matter of fact. Like weather information that's given to military commanders from weather satellites: they say, "This is really neat," but they don't understand it and they take it for granted. Everybody in the Air Force fully appreciates and understands the complexity of the planning, the execution, the secrecy involved in carrying out an operation like the Israeli raid on the Iraqi nuclear reactor. But take the space shuttle: how many people really appreciated the complexity of that system and all that had to happen to make it work right?'

The spacemen often express this feeling that 'space' is

taken for granted by the rest of the Air Force. It would be natural to ask how much of their outlook is shaped by their own career circumstances. The spacemen believe strongly in what they say, and often are making personal sacrifices to stay in the Air Force and work on space programs. (Many are trained engineers. Major Friedenstein, who organized the Air Force Academy space symposium, estimates that he could be making at least $10,000 a year more outside the Air Force.) Even so, there is a bureaucratic law that 'where you stand depends on where you sit.'

In the Air Force (as in the other services), technicians can rise to respectable ranks, but only 'leaders' rise to the very top. By definition, a military 'leader' is someone who would lead men into battle. Technicians who 'merely' support the combat units are second-class citizens in the military hierarchy. In their own biographical statements (attached to their symposium background papers), the spacemen stress, when possible, their Vietnam combat experience. They feel that if 'space' were considered an 'operational' arena of military activity, the value of their own work would be more appreciated by the rest of the Air Force, including those who make decisions on advancement. And those already at the top might lend a more sympathetic ear to the demands of the spacemen for more resources.

This fact of life leads to what may appear to the outsider as arcane doctrinal refinements. In the late fifties the Air Force adopted the term 'aerospace' to identify its medium of operations. After all, it's not really possible anyway to determine exactly where the atmosphere stops and 'space' begins – the air just gets thinner and thinner, out to infinity. Ballistic missiles go out of the air, into space, and back into the air to deliver their warheads. Space logically 'belongs' to the Air Force, the Army gets the land, and the Navy gets the sea.

But some spacemen would point out that the aerodynamic control surfaces (wings and slats and ailerons and rudders and stabilizers) don't work in space. Most real space vehicles have to be in orbit, and the physics of flying in orbit are a lot different from the physics of flying in space. Unlike airspace, outer space can't be divided into convenient theaters of operations like Europe or Southeast Asia. Space is a distinct theater of operations unto itself. If the differences between the land and the sea and the air justify having a Navy and an Army and an Air Force, then there is justification for having a separate organization for space, for treating space as a separate arena of combat.

Among the spacemen, though, there is some disagreement about how far to push both the analogy of space power with air power and the argument for a separate organization. Two participants in the Air Force Academy symposium who later published their paper wrote that '. . . in the not too distant future, space will be the dominant medium for the maintenance of national security.' Although they are Air Force officers, the two teach in the Management Department of the Naval War College. They manage to risk offending all the present services by saying that not only should there be a U.S. Space Force, but that the Space Force will be the dominant one in the future. They point out that the Army Air Corps developed just such a rationale to explain the need for an independent Air Force at the end of World War II.

Two other symposium participants, Air Force officers assigned to an intelligence analysis unit in the office of the Secretary of Defense, wrote in their paper, 'Space constitutes America's best opportunity . . . to get out in front of and shape what will eventually become the decisive arena of military competition.'

None of these four officers was working with space systems when they wrote their papers, nor did any of them seem to have an extensive career background dealing with space. They cannot be accused of letting their bureaucratic positions shape their stands. What is interesting, however, is that those whose jobs *do* relate directly to space seem to have more down-to-earth ideas about what is practical in space and what isn't.

Does Colonel Van Inwegen think space is going to become the decisive arena of combat in the future? 'There are two ways of looking at that,' he said. 'One is that you'll use it as a medium of conflict, in terms of winning and acquiring territory on the earth: I don't know about that. The other one is, though, that as things in space become more valuable in terms of *supporting* military forces, your opponent will obviously try to find a way to deny you those activities. I see space as a more *defensively* oriented conflict area, as opposed to using it for dropping nuclear weapons, or whatever.'

When the spacemen talk about the extraordinary involvement of research and development people in the operation of space systems, they are talking about the Air Force Space Division, one of several divisions of the Air Force Systems Command. The Systems Command is in charge of most of the Air Force's research and development of new weapons and equipment – everything from radios to airplanes to missiles to satellites. It oversees most of the Air Force laboratories, which conduct research on everything from aerospace medicine to Soviet military technology. What the Systems Command does *not* do is command fighting airmen, as, say, the Pacific Air Forces Command or the Strategic Air Command does. The Space Division, then, shares the stigma of the Systems Command of not being where the 'real' action is. It's more like an aerospace company than a fighting machine.

The Space Division headquarters is in El Segundo, just south of the Los Angeles airport. It also stands in the heartland of the Southern California space industry. Ten miles to the south, in Redondo Beach, lies the TRW company's Space Park. Farther north, east, and southeast are the space factories of Rockwell International. Just a mile to the north, sprawling along the Imperial Highway across from the airport, are the tan pavilions of Hughes Space and Communications. Just across El Segundo Boulevard is a two-story outpost of the TRW Defense and Space Systems Group. Across the street to the west is the six-story main office of The Aerospace Corporation.

The Los Angeles haze and California landscaping soften Space Division's edges. It looks more like a college campus than does the Air Force's real one. Inside, the warm pastels are a far cry from the depressing off-whites and grays of the Pentagon.

The reception area of Space Division could be that of any moderately sized corporation. Glass walls on three sides enclose a large fern, some begonias, and several display cases with models of the rockets and satellites that are the firm's hardware. Along a wall past the guard's station stands a trophy case; above that, the motto, 'The Peace of the Future Is Our Profession' (a variation on the Strategic Air Command's 'Peace Is Our Profession'). Office walls sport color photographs of company products: pictures of the earth from space, rocket launches, satellites.

In a Public Affairs Office briefing room, Captain Linda Wyatt described the organization of Space Division, its products and services, its contribution to the national security. 'This fiscal year,' she said, 'we will expend approximately three billion dollars in development and acquisition and in accomplishing our missions. General [Richard] Henry has the responsibility for commanding twenty-two thousand personnel. Of those, four thousand are military, three thousand civilians, over thirteen hundred members of Technical Staff – employees of Aerospace Corporation, and over thirteen thousand contractor personnel.'

Back at the Pentagon, Colonel Van Inwegen had mentioned the considerable dependence of the military space systems on private companies for their day-to-day operation. He said, 'I think we do need less dependence on contractors . . . we have contractors flying our airplanes, for example. It creates a kind of vicious circle: if you have your satellites and your space systems operated by contractors, and you have a small number of military people involved in that, they look to the guys next to them making two or three times as much money . . . you're training people for the civilian world. But if you had more job opportunities in the military, you'd have a base line of people. Not only do we want to retain military people, but as we build more systems that are more directly involved in military operations, we want military people running them.'

At Space Division headquarters, where they *don't*

actually operate the space systems, there is an unusually high proportion of uniformed military people to civilians – but still only about 20 percent of 5,800 employees. Another 900 are civilian government workers, but 350 more are contractor people and 1,200 work for The Aerospace Corporation. At headquarters they 'manage programs' – they oversee the contractors who develop and build the satellites, rockets, other hardware, and computer programs that make up the space systems.

It takes skilled scientists and engineers to manage such technical programs. The Air Force has some of those people, but not nearly enough. There are several obstacles to getting them. For one thing, there is the military tradition of moving young officers around, giving them a wide variety of experiences as they move up the ranks; yet, an officer who manages to stay in one place too long may have a harder time getting promoted. All the same, you need specialists and their years of experience to oversee such technically specialized programs.

Another problem is money, and it's getting worse. Lieutenant General Richard Henry, Space Division commander, says, 'My loss rate of middle management is twice the Air Force average because people cannot afford to come out here. I cannot recruit civilian or military personnel. As a result, I have a very strong force of second lieutenants who are very, very good, but inexperienced. We are rapidly losing the corporate knowledge and experience we used to have.'

Henry is trying to hire civilians as well. A Space Division ad in the *Washington Post* reads: 'Your Space Career Can Really Take Off in California . . . Our journey into space creates a work environment charged with excitement and challenge for Industrial Engineers at GS-12 and GS-13 ($26,951 to $41,660) and Contract Negotiators and Price Analysts at GS-11 and GS-12 ($22,486 to $35,033).'

In fact, it was just to supply such technical manpower that the Air Force originally set up The Aerospace Corporation. In the late 1950s, when the Air Force was developing the Intercontinental Ballistic Missile, it hired the Ramo-Wooldrige Corporation (in 1958 it became TRW) to do the 'systems engineering.' Roughly, that means organizing the building and installation of a complicated new piece of hardware. TRW's competitors began to complain. Although TRW might be banned by contract from supplying the hardware for the missiles it was helping to design, it still could get unfair access to the industrial secrets of its competitors.

In 1959 a subcommittee of the House of Representatives Committee on Government Operations recommended that the division of TRW involved, the Space Technology Laboratories, '. . . be converted into a nonprofit institution akin to the Rank Corporation and other private and university-sponsored organizations which serve the military departments and other agencies of the Federal Government on a stable and continuing basis.' So in 1960 the Air Force bought the Space Technology Laboratories from TRW. It chartered the new nonprofit Aerospace Corporation, whose building and 75 percent of whose people came from STL. (For some years, though, TRW retained a division called STL.)

Although Aerospace has had some contracts with other government agencies – mostly NASA and the Department of Energy – by far the bulk of its revenue still comes from the Air Force: $161 million in fiscal year 1980 and $197 million in 1981. General Henry may be having trouble hiring people, but Aerospace does not seem to be: in 1980 it added 400 employees, making a total of more than 4,000.

The corporation is one of four Air Force Federal Contract Research Centers – a device used by other government agencies as well. This scheme allows the government to hire highly skilled scientific and engineering manpower at more or less competitive rates without appearing to add to the total number of civil servants. The companies are semiautonomous, but for all practical purposes the workers are government employees. At Space Division headquarters (and elsewhere in Space Division units), the civilians work side by side, in the same offices and control rooms, with the blue-suiters.

Van Inwegen had mentioned the high proportion of contractor personnel in space programs. He had also mentioned the continuing emphasis on research and development. According to General Henry, 'What we have not been able to do in the past twenty-three years is draw a hard line between research and development and production. Every spacecraft tends to be different. The classic production that we associate with airplanes or ships has not prevailed here.

'The other thing we have not learned how to do is define the demarcation line between R and D and operation. The largest visible evidence of that is in my mission [of both overseeing research and operating the satellites]. The fact that we have not been able to transition the launch of a spacecraft, or the tracking and control of a spacecraft, to operating command reflects the nature of the business, perhaps the infancy of the business. We have not been able to figure out a logical way to transition into what we would call space operations.'

In November 1971, the Air Force launched the first two of a series of communications satellites into 23,000-mile-high orbits. At first, one refused to accept radioed commands from the ground. They couldn't even find the second. It took four days to find them both and get them both working.

These satellites, built by TRW, maintained their stability – were kept from gyrating randomly – by spinning, like a gyroscope. But to keep the antennae pointing toward the same spot on earth, the center section of the satellites had to be 'despun' with an electric motor. After working for a month, the two satellites started rotating uncontrollably, pointing their antennae away from the earth. The ground controllers managed to straighten one out, breaking an antennae hinge in the process. The multimillion-dollar satellite worked, but didn't give its money's worth.

TRW and Aerospace engineers went to work on the other satellite. They simulated the characteristics of the satellite and its orbit on Aerospace's big computer in El Segundo. They figured out that the despinning motor was pulsing in such a way as to swing the satellite out of position. Then they used the computer to simulate a satellite command center, on which they practiced sending instructions to the satellite that would correct the spinning problem. After some practice, the engineers, now trained 'satellite pilots,' telephoned orders to the real satellite controller, who was then able to set the satellite straight. Later satellites in the series had to be redesigned to avoid the problems of the first two.

Aerospace Corporation engineers came to the rescue again in 1976 when the first of a new series of military weather observation satellites started tumbling in space. Not being able to keep its solar power collecter aimed at the sun, the satellite lost its battery power. Later RCA, the builder of the satellite, figured out how to get some power back and got the satellite under some control – but not enough control for it to take its weather pictures. Once again, Aerospace set up a simulation on its computer. The data in the computer copied the actions of the satellite, but went throught those actions two thousand times faster than the actual satellite. Commands to the satellite were tried out on the computer, then on the satellite, then on the computer – back and forth until the satellite was brought under control. The Air Force, Aerospace, RCA, and other labs brought in on the job took six months to get the weather satellite delivering the goods. And then it really only worked properly for another month.

The second weather satellite in that series came up with similar problems. The third and fourth had other difficulties. The launch of the fifth one, in 1980, failed. The sixth went up in 1981. Now, the next generation of weather satellites is on the way. You can begin to see why Van Inwegen and other Air Force 'operations' people complain about the heavy role of R and D. But with the continuing 'technological enamorment' of the space program planners, the R-and-D guys are going to have to be there to keep the satellites working.

Space Division is responsible for the design, building, and launching of military satellites, just as the other divisions of the Air Force Systems Command are responsible for producing their brands of hardware. But unlike the other Systems Command divisions, Space Division is also responsible for running much of the hardware it supplies.

Headquarters for most Department of Defense satellites are in the Big Blue Cube.

In Sunnyvale, California, just off Lockheed Way, is the command center of the Space Division's Satellite Control Facility. The Satellite Test Center, as these headquarters are known, is a nine-story, windowless, pale-blue block, with an attached administrative building and several massive white radar and communications dishes in the parking lot. There are also seven remote tracking stations in the Facility, but these have limited abilities to send commands up to the satellites. The Satellite Test Center is the brain of the system.

The Big Blue Cube – in fact the whole complex of stations that make up the Satellite Control Facility – is dominated even more by contractor employees than Space Division headquarters. The Satellite Test Center is in Sunnyvale because it is close to Lockheed, whose employees run the mission control centers. In 1981, there were 855 Air Force employees at Sunnyvale: 226 officers, 399 enlisted men, and 230 civilians. There were 58 Aerospace Corporation workers. But there were 926 contract employees, mostly from Lockheed.

At the seven other Satellite Control Facility stations around the world, there are about 750 Air Force employees, and about 950 contractors, mostly from Ford Aerospace and Communications. The locations of these Remote Tracking Stations range from lush to desolate. The most exotic is the Indian Ocean Station, 1,000 miles east of Kenya, in the Seychelles. The U.S. Air Force base on Guam, in the South Pacific, has a station. There's another in Oahu, Hawaii. (The Satellite Control Facility runs another operation in Hawaii: the 6594th Test Group. This group has the unique mission of making midair catches of film capsules ejected from reconnaissance satellites.) Another is near Thule, Greenland, about 800 miles from the North Pole. The Air Force shares a tracking station in England with the Royal Air Force: Oakhanger, southwest of London. (Ford built the United Kingdom's Skynet military communications satellite, but doesn't run the tracking station.) There are two tracking stations in the continental U.S.: one near Manchester, New Hampshire, and the other at Vandenberg Air Force Base in California.

The Satellite Control Facility – these seven Remote Tracking Stations plus the Satellite Test Center in the Big Blue Cube – operates about 45 U.S. military satellites. All together, the various stations and the Center make contacts with satellites about 90,000 times a year (that averages out to about five contacts per satellite every day). When the SCF workers make a scheduled contact with a satellite, they carry out what they call a 'pass plan.' They keep track of the 'ephemerides' – the orbital path – of the satellite. They read out the telemetry (the transmitted data) from the satellite, which tells them the condition of the satellite, and it also provides the information (weather, intelligence, etc.) that the military customers use. Finally, the satellite controllers send instructions to the satellite, perhaps telling it to fire its thrusters to change or maintain its orbit, or to turn a camera on or off, or to radio down the information it has gathered.

There's no such thing as a visitor's tour at the Satellite Test Center. There's no briefing for the press. In fact, unlike most Air Force bases, there's no Public Affairs Office. There is a cheerful reception area, with wood-paneled walls, potted plants, and colorful rocket and satellite pictures. But past the guard's desk, a sign announces: IDENTIFICATION REQUIRED BEYOND THIS POINT/USE OF DEADLY FORCE AUTHORIZED.

Inside the Big Blue Cube are seven mission control centers, each assigned its own type of satellite, its own computers, and its own channel of communications through the tracking sites to the satellites it runs. Five of the Remote Tracking Stations can maintain contact (carry out pass plans) with two satellites at once, meaning that the system as a whole can communicate with up to twelve satellites at any given time.

The Space Division is planning for a heavier work load in the future. First there is the Modernized Remote Tracking Stations plan (MRTS). Modernizing the Remote Tracking Stations in this case means more heavily computerizing and centralizing the system (IBM has a $92 million contract for the job). Computers at the Control Facility headquarters will automatically prepare the remote stations for a satellite pass in five minutes' time. Less of the raw data will be processed at the tracking stations, more at the Control Facility with its latest high-speed computers. Space Division expects that these changes will let them cut the staffs of the Remote Tracking Stations (all those Ford people) by half.

Meanwhile, at Satellite Control Facility headquarters, the Data Systems Modernization project will place more and more of the communications between ground and satellites under computer control. Computers will manage the scheduling of satellite pass-plan executions and the switching of the equipment from one operation to another. Computers will analyze more of the telemetry from the satellites, analyzing, storing, and displaying the data with little human intervention.

By the end of 1987 the Space Division is to have built an entire new Consolidated Space Operations Center. This center, near Colorado Springs, will become part of the new Air Force Space Command. It will handle new satellites in addition to the ones piloted out of the Satellite Test Center in Sunnyvale. It will also serve as a backup, able to operate the STC's satellites as well, in case an earthquake or a terrorist attack should knock out the STC. The consolidated center will not only run many of the Defense Department's unmanned satellites, it will also house the Air Force's Shuttle Operations and Planning Center. Until that center is working, the Air Force will run its space shuttle flights out of its own control room at NASA's Manned Space Flight Center in Houston.

The Space Division is also at work on projects that might allow it to get rid of its foreign Remote Tracking Stations. One such idea is the Satellite Control Satellite, which would relay information back and forth between military satellites and ground stations only in the U.S. NASA is already building something like this in the TDRSS – the Tracking and Data Relay Satellite System – whose four spacecraft will go up on the space shuttle in 1983 and 1984. The Air Force may make some use of the TDRSS, but that system has some disadvantages from the Air Force point of view. For one thing, the first two TDRSS satellites probably won't have the data encryption equipment the Air Force wants for secrecy. Second, the satellites will only be leased by NASA from Western Union: the Air Force would rather have satellites it designs and runs itself.

Another way of making the Satellite Control Facility less dependent on foreign tracking stations is to make the satellites less dependent on the ground. General Henry looks to microcomputers to make satellites more self-sufficient: 'Today, spacecraft have an umbilical link with the earth for telemetry tracking and control – it's too big an umbilical link. Spacecraft could do a number of things if they had onboard computation for stored logic. For instance, when a communication channel goes down, the spacecraft could move to another channel, or it could go to a backup mode for position, or it might exercise its thrusters to move into another orbit. I think autonomous spacecraft should be the wave of the future. As such, they would essentially be self-healing and self-navigating . . . [with the umbilical link gone] we can be rid of that additional overhead and have more survivable spacecraft that are harder to find and destroy, and which will be more reliable.'

Space Division has given Martin Marietta Aerospace in Denver, Colorado, a contract to design and build a space sextant, the Autonomous Navigation and Attitude Reference System. (This instrument got its first space test aboard the fourth flight of the shuttle at the end of June 1982.) On board a satellite, the space sextant will use two telescopes and a computer to measure the angles between the moon and any of twenty-four stars. It'll measure the position of the satellite in space faster and more accurately than ground stations can now. It will keep the attitude of the satellite – the direction in which it and its instruments point – within one second of arc, which is more accurate than ever possible before. Most importantly, the space sextant will make satellites far less dependent on ground stations. That means that the Soviets would find it much harder to interfere with the satellite with electronic signals. And it means that the satellite could keep operating even if a ground station were out of service.

Although the Space Division runs and passes along the data from most U.S. military satellites, it doesn't do so for all of them. One of the complaints of the spacemen has been that there's considerable fragmentation of control over the military uses of space. For example, until recently the Strategic Air Command, not Space Division, ran the two Defense weather satellites. The control center, at Offutt Air Force Base, near Omaha, now answers to the new Space Command headquarters in Colorado Springs. The two main 'command readout' sites, where the weather satellites dump their stored information, are at Fairchild Air Force Base in Washington State and at Loring Air Force Base in Maine.

The U.S. Navy has its own space program. As the Satellite Test Center it has its own mission control team for its Fleet Satellite Communications System satellites. (It probably uses the same facility to run its highly secret ocean surveillance satellites.) The Naval Astronautics Group at Point Mugu, California, runs its Transit navigation satellites. In October 1983, the Navy formed its own Space Command. The Navy resisted Air Force suggestions for a Unified Space Command over all military space activities. It may be willing to cooperate up to a point in Air Force efforts to centralize space management, but it also wants to control its own satellite resources. It wants to make sure that it will keep getting the support it needs from the Air Force to keep its satellites running. And it wants to be sure that that when the U.S. anti-satellite system comes on line, the Number One target will be the Soviet ocean surveillance satellites that could direct missiles and bombers to Navy ships.

Thanks to a notorious spy trial, we now know of yet another kind of satellite that Space Division doesn't run. This satellite, in an orbit that keeps it in one spot over the earth's equator, detects the telemetry (radioed data) from missile tests inside the Soviet Union. The main control and readout station for the satellite is at Pine Gap, Australia. In 1977 a young TRW employee, Christopher Boyce, was sentenced to forty years in jail for selling information about the satellite to the Russians. That's how we know that the data picked up at the CIA's Australian base is processed at TRW's Space Park in Redondo beach before it goes to the CIA headquarters outside Washington.

The Aerospace Defense Command (ADCOM), headquartered in Colorado Springs, Colorado, takes data directly from three early warning satellites also in stationary orbits over the equator. These satellites detect rocket exhaust plumes; they'd be the first source of warning the U.S. would have that Soviet missiles were on the way. One readout station is in Australia (it has to beam its information take back via other satellites or by cables) and the other is at Buckley Air National Guard Base outside Denver.

The Aerospace Defence Command also has charge of the Space Defense Operations Center (SPADOC). Until recently, here's where the most significant split in U.S. military space authority lay. The Space Division's Consolidated Space Operations Center will be a few miles from the Aerospace Defense Command's Space Defense Operations Center, but until September 1982, the two were to have been under entirely different military commands.

Now those commands will overlap within the Space Command. The general in charge of ADCOM will also head the Space Command. The general heading the Space *Division* (of the Systems Command) will be the vice commander of the Space Command. And *his* deputy will head the Consolidated Space Operations Center.

By virtue of its connection with the Aerospace Defence Command, the Space Defence Operations Center lies 1,700 feet inside a granite mountain. It's part of the NORAD (North American Aerospace Defense Command – so called because it includes Canadian units; the U.S. component is the Aerospace Defense Command) Cheyenne Mountain Complex. Back in 1956 Air Force air defense generals decided they needed an underground command post that would be safe from Soviet H-bombs. Since the North American Air Defense Command (as it was then called) was already headquartered at Colorado Springs, they looked around that area for a suitable site.

Construction contractors started carving out chambers inside Cheyenne Mountain in 1961; NORAD moved in in 1966. Given the capabilities of modern Soviet ICBMs, nobody's counting anymore on the survival of the command post in a nuclear war. Its main job now is to be a warning center.

NORAD's original job was to look out for Soviet bombers and send out jet interceptors to meet them if need be. Now its main job is merely to report to the President, the Pentagon, and the Strategic Air Command if Soviet missiles are coming our way. But in looking out for Soviet missiles, the Aerospace Defense Command needs to look spaceward. It was logical that it should take command of the Air Force's Space Detection and Tracking System (SPADATS).

The NORAD Space Defense Computation Center is plugged into nineteen radar stations and nine special telescopic cameras around the world. These include Ballistic Missile Early Warning System radars in England, Greenland, and Alaska. They include four large radar bases in the continental U.S. whose main job is to look out for incoming missiles. They include two radar stations in the Aleutians and Turkey whose main task is to monitor Soviet missile tests. They include U.S. Navy Space Surveillance System radars in the Atlantic (on Ascension and Antigua islands), and one in Hawaii. The special cameras are in New Zealand, Korea, Hawaii, California, New Mexico, Canada, and Italy.

At any given time the SPADOC Computation Center's computers are keeping track of the orbits of about 4,500 man-made objects in space. One of SPADOC's jobs is to try to predict when satellites will come down. For example, when the Russian nuclear-powered radar satellite Cosmos 954 fell on Canada early in 1978, the U.S. government was already prepared to send teams to the Northwest Territories to help clean up the radioactive mess.

But SPADOC is interested in *any* change in Soviet satellites. The space watchers keep track of the maneuvers of spy satellites – and help other American military units hide their own secret activities from those eyes in the sky.

They look out for tests of the Soviet anti-satellite satellite. They're ready to report if it seems that the Soviets are attacking U.S. satellites. And they're now gearing up to manage their own combat operations – to be the command post for the new U.S. anti-satellite weapon that the Air Force plans to test in 1983.

I visited the man who had been in charge of converting a mere Space Computation Center into the Space Defense Operations Center in 1979, and then became SPADOC's first director. Colonel Richard Schehr, USAF (ret.), a 1960 Air Force Academy graduate, was a classmate of Colonel Van Inwegen's. After his retirement from the Air Force early in 1981, Schehr went to work for the Systems Development Corporation's Colorado Springs branch.

The SDC has done considerable work at the Cheyenne Mountain Complex, mostly in preparing the 'software,' or programs, for the large Honeywell and Philco-Ford computers that track the satellites and missiles. (Ford Aerospace has done work for the Aerospace Defense Command as well as for Space Division. Ford has been the prime contractor for the Cheyenne Mountain communications, computer, and display systems, as well as maintaining the computer programs permanently – it's gotten at least $90 million in contracts at Cheyenne Mountain.)

Schehr, then, is a longtime 'space guy' – or, as he puts it, one of the 'space Mafia.' He prepared one of the background papers for the Air Force Academy space symposium. Schehr had argued for a closer link between the new Consolidated Space Operations Center (CSOC) planned for Colorado Springs and the Aerospace Defense Command (ADCOM).

This link, Schehr had prophesied, 'would start to pull a Space Command together. It would also mean that now a four-star general [the head of ADCOM] would start to speak for space – he would have a vested interest in its well-being. And the three-stars on the Air Staff would have to listen; one three-star isn't going to listen to another, but they always listen to fours.'

Schehr thought that at least the Space Defense Operations Center could be the place where all U.S. military space activities start to be thought about under one organizational roof – a locus of information and potential coordination among the various space operators. 'It would begin to pull all the military resources of this country together,' he said. 'It's very important that people begin to understand all aspects of space. It's true that secrecy does limit understanding. The problem is that when you don't have it under one command, the right hand doesn't know what the left hand is doing; sometimes you waste money. SPADOC offers the best chance of pulling all the information together. It can be the "librarian" of space, if you will – not the owner, not the controller, but the information central. The place where at all classification levels the information comes together.'

And why bring it all together? So that the operators of the satellites, the space watchers, and the commanders of the coming U.S. anti-satellite system can work out a unified space battle plan. They can be prepared to defend American satellites and attack Russian ones. The spacemen will have transformed themselves then from mere supporters of the main action into real combat officers.

# CHAPTER TWO

# The Space Business

Five rocket engines, 6.5 million pounds of thrust, heave a 235-ton glider into space; after several 18,000-mile-an-hour circuits of the earth, the crew flies it back. The bird drops like a brick out of the sky and rolls straight down the center line of the landing strip. We cheer.

Maybe the thrill we have for the space shuttle has something to do with the 'ours-always-blow-up' syndrome Tom Wolfe describes in *The Right Stuff*. At least for the first few times we have to wonder if the brand-new spaceship is going to make it up and back. But it may have more to do with the feelings we had when Neil Armstrong first bounced onto the moon's dust. We had set out to do something nobody else had done before, and we made it.

Still, there is irony in the accomplishment. The space shuttle was supposed to be different from the Apollo project. Going to the moon was a sort of twentieth-century Great Pyramid. NASA made a lot of the 'scientific' experiments and the 'spin-offs' of exotic new technologies, but the main thing had been to show that we could get there, to show ourselves, the world, and the Russians – most especially the Russians – how clever and capable we are. Pyramid projects are not necessarily boondoggles: the rewards in national pride can be rich. National pride, particularly when subliminated into nonmartial pursuits, is worth having.

The National Aeronautics and Space Administration sold the space shuttle to Congress on more down-to-earth grounds. It was to be more of a federal highway project, or a national railroad land subsidy, than a pyramid. Going to the moon was an end in itself; building the space shuttle was to be an economic infrastructure project. It was to be *the* National Space Transportation System (as the shuttle is officially known), our truck in the space lanes. It would make it cheaper and easier to get scientific experiments, weather and communications satellites, and military payloads into orbit. Much cheaper, said NASA, because we won't be throwing the rocket away every time we send one up.

It's not that NASA hadn't wanted another Great Pyramid. After the successful moon landings, NASA's dreamers envisioned a twelve-man space station (to be followed by a hundred-man space base), space stations on the moon and in orbit around it, and a space shuttle to help everybody get there. In 1969 and 1970, however, we had the Vietnam War to pay for. NASA gave up on the moon and decided to settle for a space station going around the earth and for the shuttle. Still too expensive. NASA would have preferred to settle for just the space station, but its analysts figured out that without a cheap way of getting people and equipment back and forth, the station would still cost too much.

In 1970 NASA thought it could get a fully reusable two-stage space shuttle for about $5.2 billion in development costs. The first, or booster, stage would be about the size of a Boeing 747. It would carry the second stage, about the size of a 707, to high altitude, then fly back to earth, landing like a plane. The second stage would go on into orbit, hauling a payload of 25,000 pounds. NASA gave both McDonnell Douglas and North American Rockwell contracts to design such a shuttle.

While those studies were going on in 1970, NASA figured out something else: if it was ever going to sell this project to Congress, it was going to have to sell it to the Pentagon first. In January 1971, NASA incorporated Pentagon specifications into its shuttle plans. The bird would have to carry 65,000 pounds into orbit, not just the 40,000 that its latest designs called for. It would have to be able to land anywhere within a 1,700-mile-wide swath.* The addi-

---

* The Air Force wanted this 'cross-range' so that it could send the orbiter on a single-orbit reconnaissance mission. If the shuttle took off from Vandenberg Air Force Base and went once around the poles, by the time it returned to its starting point, the earth below would have turned to the east by about 25 degrees. The orbiter would have to maneuver hundreds of miles through the atmosphere to get back to base.

tional power needed raised the estimated development cost to almost $10 billion.

The Defense Department (and its primary agent for space, the Air Force) did not intend to pay for developing the shuttle. They said, 'If we're going to have to use your shuttle, this is what we want it to do.' And implicitly, 'If you want us to go to bat for you before the Congressional committees, you'd better make sure we get what we want.' If NASA was going to sell the shuttle as the National Space Transportation System, it needed the Pentagon; the Pentagon didn't need it. But with or without the Pentagon on board, NASA was not going to get $10 billion out of a Congress that was financing a $30-billion-a-year war in Indochina, and that was a little jaded with space.

NASA went back to the drawing boards. The new shuttle would only be half reusable: it would consist of an orbiting vehicle with three main rocket engines, a large external fuel tank that would be thrown away, and two extra solid-fuel rocket boosters that would fall into the ocean to be recovered and used again. This is the plan that President Richard Nixon approved in January 1972. NASA guessed the cost would be $5.5 billion. By the end of 1972, it had that figure down to $5.15 billion, including the first two orbiters and the first six flights (extra orbiters would be $250 million apiece).

NASA got bids for its compromise space shuttle from four companies: Grumman Aerospace, Lockheed Missiles and Space, McDonnell Douglas, and North American Rockwell (now the North American Space Operations Division of Rockwell International). Rockwell came in a billion dollars lower than everybody else: it would develop the orbiter, the spaceship itself, for $2.6 billion in six years. Six and a half years later, and still two and a half years from the first test launch, the orbiter development cost was $5.2 billion and mounting. Rockwell's Rocketdyne Division also got the contract for the space shuttle main engines, separate from the orbiter contract. Original estimate for the engines: $580 million. Estimate for the main engines through fiscal year 1980: $1.17 billion. The estimate in mid-1979 for total shuttle development costs had risen to

$8.4 billion. By 1980, the General Accounting Office was projecting upward of $13 billion.

Few would now deny that in 1972 NASA and Rockwell 'bought in' on the shuttle. 'Buying in' is a classic defense contracting ploy: you go to Congress with a stripped-down, bare-essential, everything-must-go-perfectly cost estimate. Congress says, well, okay. Then reality moves in. Engineers' plans have to be turned into working hardware. Nobody has ever built a rocket engine that packed so much power into so little space, and could be used over and over again, besides. Nobody has ever built a giant space capsule that could fly back to earth like a glider, heating up to 2,600 degrees Fahrenheit on the outside and staying cool on the inside.

But the budget estimates contained no fudge factors. Everything would have had to have worked right the first time. That wouldn't have been just good management: that would have been a miracle. So NASA and Rockwell bought in. Defenders of the cost overruns like to point out that if you take inflation into account, the shuttle development program really comes in at only about 20 percent above the original estimates. But of course, the shuttle builders bought in on time as well as money. If you add, modestly, three and a half years of delays, that's another 25 percent added by inflation that came from buying in. And each additional shuttle orbiter now costs a billion dollars, which, even counting inflation, ups the ante considerably from the original estimates of $250 million apiece.

When NASA bought in, one of the things it may not have figured on was the recalcitrance Congress and successive Presidents would show in paying the extra bills. Bare minimum financing every year further stretched out the program, raising the ultimate costs even higher. In 1973, NASA estimated that seven orbiters would fly 725 missions. By 1976, NASA and the Defense Department agreed that between them they would need an absolute minimum of five orbiters. Both the Ford and Carter administrations delayed buying the fifth orbiter. In 1978, the General Accounting Office (an agent of Congress) recommended that there should only be three. Since then, the option has remained open on the fifth.

Meanwhile, the Space Transportation System is still a long way from providing the kind of service that NASA promised it would. The first few shuttles will not be able to put the specified 65,000 pounds into orbit. Rockwell (and Martin Marietta, which makes the giant external fuel tank) engineers are still trying to figure out how to reduce weights and increase power to meet the Air Force payload requirements. NASA planners had figured on being able to relaunch ('turn around') any given orbiter within two weeks of its last trip into space. Now they think they'll be lucky if it's four to six weeks. As late as the beginning of 1980, NASA had scheduled 109 operational shuttle flights for the years 1982 to 1986. By mid-1981, the number of flights planned for that period was down to 62. Thirty to forty percent of the launches would be military.

The fact is that if it weren't for the Defense Department – specifically, the Air Force – the shuttle would be dead. In 1980 NASA needed an extra $300 million just to keep the project alive. President Jimmy Carter was not too sympathetic until the Defense Department told him how necessary the shuttle had become. Congress would surely not have voted the extra money over the last few years if the Defense Department hadn't gone to bat, saying how much they needed it.

There's some debate, though, about who's doing whom a favor. Space scientists, for example, had hoped that the end of the Apollo moonshots would finally free up NASA money for more scientific research in space. Instead, the shuttle has eaten up more and more of a shrinking NASA budget. Planetary scientist Carl Sagan says: 'The space shuttle is to a significant degree pumped by military requirements. Whether that's good or bad is a separate issue, but it seems to me this is a case or borrowing – or stealing – from the poor, namely NASA, to benefit the rich, namely the Department of Defense. The NASA budget is a very tiny fraction of the Department of Defense budget, and yet it is NASA that has to develop the launch system for Department of Defense reconnaissance and other satellites. . . . This country has very few places where we are acknowledged world leaders, especially in activities that everybody says benefit the entire population of the planet. To take away from that very small pot of money and give it to the Department of Defense, which has such an enormous budget – over a trillion dollars for the next few years – seems to me a serious mistake.' (The General Accounting Office has estimated that 20 percent of NASA's 1983 budget should be counted as supporting the Department of Defense.)

It's true that the Air Force is not paying directly for the costs of developing the shuttle. But it is putting up a few billion to facilitate its own use of the shuttle. In fact, the biggest single piece of Space Division's budget recently has gone to getting the Air Force ready for the shuttle. Several shuttle-related projects will cost the Air Force $3 to $4 billion, depending on what you count and how you count it. NASA and the Air Force cut a deal whereby NASA would prepare Cape Canaveral for shuttle launches and the Air Force would ready a launch pad at Vandenberg Air Force Base in California; that project will probably end up costing the Air Force about $2 billion. The Air Force also contracted with Boeing Aerospace to build an upper rocket stage – a vehicle to carry satellites from the low orbits of the shuttle to higher orbits. Latest cost estimates top $700 million. NASA will use the Vandenberg launch pad and the upper rocket stage, though whether it really needed them is another question.

Since the Air Force wants to control its own shuttle flights, and to do so in secrecy, it's putting up another $250 million to build its own command post at the Johnson Space Flight Center in Houston, to tighten up security and communications secrecy at Cape Canaveral, and to put in computers at NASA's communications central, Goddard Space Flight Center, so that transmissions to and from the shuttle can be encrypted.

Vandenberg Air Force Base stretches for about 25 miles along the coast of the southwesternmost corner of North America. The geography is not irrelevant. In 1978 the General Accounting Office told Congress that NASA and the Air Force could save the tax-payers about $3 billion by launching space shuttles only from Florida. The Air Force and NASA replied that they *had* to have the California launch site.

Rockets from Cape Canaveral fly first over the Atlantic Ocean; if something goes wrong, they'll probably ditch there without hurting anything. But that only works if the rockets are aimed more or less due east. Point them much to the south and their early trajectory takes them over the Caribbean islands and at least part of South America. Point them much to the north and they're passing along the Eastern Seaboard and into Canada. (What's more, any satellite launched from the U.S. in a northerly direction will be taking the most direct route to the Soviet Union, something to be avoided if you don't want the Soviets to mistake your satellite launch for a nuclear missile.) If you want a satellite's orbit to pass north of 56 degrees latitude, you have to do one of two things. The first thing is to send the launching rocket on a 'dogleg' – first east into the Atlantic and north or south when it gets safely away. This option uses up energy. Precious rocket fuel is used in changing the rocket's course. That loss inexorably reduces the rocket's payload – the weight of the satellite delivered into orbit.

The second way to get a satellite into an orbit with a relatively high angle of inclination to the equator is to launch it from the southwest coast. Launches from there can go on in whatever southerly direction is necessary to get the satellite's orbit more perpendicular to the equator, closer to the poles. Such polar launches are more costly in payload and fuel anyway, because the rocket can't take advantage of momentum it has from the rotation of the earth when it travels west to east. But at least the fuel penalty of the 'dogleg' launch is avoided.

NASA and some of its customers find these polar or higher inclination satellite orbits useful. The Tiros and Nimbus series of weather satellites use them, and so does the Landsat earth resources satellite. But the Defense Department finds the higher inclination orbits essential. In order to cover the whole earth, picture-taking reconnaissance satellites have to pass more or less over the poles. Important Air Force Communications satellites, Air Force weather satellites, and Navy ocean reconnaissance satellites use orbits at higher angles of inclination to the equator than the 56 degrees attainable from Cape Canaveral. Of the first ten space shuttle launches planned from Vandenberg, one will have a NASA payload, one will be a 'makeup' mission for either NASA or the Defense Department, and *eight* will carry military satellites or experiments.

The space shuttle as Great Pyramid comes to mind when you see 'Slick Six.' Space Launch Complex Six – SLC-6 – is a monumental construction project. In fact, it's a monument to an earlier, dead space project. In the 1960s the Air Force spent $1.5 billion preparing to launch its Manned Orbiting Laboratory (MOL), before the project was canceled. Slick Six was built for the MOL, then mothballed. The space shuttle launch pad is designed to take advantage of the facilities built for the MOL. The pad's most impressive feature is the Mobile Service Tower, a giant three-sided building on wheels. Its roof has been lowered by 40 feet, so it's now 275 feet high, 135 feet long, 85 feet wide, and it's lost some of its original 7,700 ton weight.

Slick Six lies in a U-shaped Valley, its back to the foothills of the Santa Ynez Mountains, its face to the rocky Pacific Coast. In order to allow the Mobile Service Tower to roll back 450 feet when the space shuttle takes off, construction workers had to gouge another 150 feet out of the side of the hill against which the complex rests. Before the shuttle is launched, though, the Service Tower will sit right over the pad: its 200-ton crane will help stack up the solid rocket boosters, the 154-foot liquid-fuel tank, and the 122-foot orbiter.

On the opposite side of the launch pad will stand another mobile building, the Payload Changeout Room. This building will first haul the external liquid-fuel tank to the pad, then the orbiter. Finally, it will make a third trip up to the pad to install the shuttle payload.

Even in its raw state – steel frames and concrete masses, exposed earth and construction-site clutter – the great spaceport is awesome. As I overlooked the complex from one of its enclosing hillsides, the monster Mobile Service Structure was to my right. To the left was the skeleton of one of the fixed buildings, the Payload Preparation Room, where shuttle payloads will be stacked in the vertical, ready for the Payload Changeout Room to haul them to the pad. Directly below was the launch pad itself, where its 100-foot-long Launch Mount for the shuttle and the 263-foot-high Access Tower were under construction. On the hillside to my right (and to the rear of the Mobile Service Structure) were water tanks that will feed a million gallons of water a minute to the launch pad when the shuttle takes off. The water absorbs the enormous acoustic energy generated by the rockets that would otherwise wreck the shuttle and the launch pad. Not yet begun was the $40 million movable enclosure that will be needed to shield the loaded shuttle from the high winds that often blow in from the Pacific.

Below the launch pad, workers were constructing two exhaust-flame ducts in addition to the one already built for the MOL project. Each duct is 50 feet deep, 70 feet wide, and has 9- to 12-foot-thick walls. These ducts will use 87,000 cubic yards of concrete, enough, Air Force public officials point out, 'to build nearly 450 miles of three-foot-wide, four-inch thick sidewalk, or, for example, a sidewalk from Los Angeles to San Francisco.'

The whole project, they point out, will take 'nearly 250,000 yards of concrete, the equivalent of a 25-mile four lane interstate. It also requires 9,000 tons of steel reinforcing bar and 15,000 tons of structural steel, the equivalent of a 120-story office building.' Other modifications of Vandenberg included in the package are the lengthening of its 8,000-foot runway to 15,000 feet to accommodate shuttle landings, building a hangar for servicing the orbiters, installing computers and other ground equipment for running shuttle launches, and altering an old Coast Guard life-boat station to accept transfer of the 154-foot external fuel tank (the tanks will come in by barge from New Orleans via the Panama Canal).

Like NASA, though for different reasons, the Air Force has had trouble delivering its end of the shuttle project on time and on cost. Slick Six was originally supposed to be ready to go in December 1982. That target date has repeatedly been postponed – first to June 1983, then to December 1983, then to June 1984, then to October 1984, finally to October 1985. The launch site was originally supposed to accommodate twenty shuttle launches a year; now it is being prepared to handle only ten. As late as 1977 the Air Force thought the project would cost about $798 million. By 1980 the estimate had risen above $1.8 billion. It's still rising.

The Air Force has also had problems with its other major shuttle-related project, the Inertial Upper Stage, or IUS. The space shuttle will only carry satellites into orbits a few hundred miles above the earth's surface. Satellites destined for higher orbits, particularly the earth-stationary 22,250-mile orbit, will have to ride on still another rocket stage. The Air Force's IUS is the stepchild of yet another disappointed NASA ambition. When NASA planners began work on the shuttle, they proposed that there should be an 'orbital transfer vehicle' – a space tugboat that would travel back and forth between the lower shuttle orbiter altitudes and the higher altitudes of such things as communications, weather, and early missile warning satellites.

Until this space tug became available, though, an 'interim' upper rocket stage would be needed. Unlike the space tug, this IUS would be expendable – a one-time vehicle that did not return from the higher orbits for another load as the tug would. The Air Force agreed to pay for the development of this short-run substitute for the space tug. Space Division hired Boeing Aerospace to do the job. As shuttle costs mounted, though, NASA kept putting off the space tug idea, and finally gave up on it. Almost before anyone had noticed, the 'Interim' Upper Stage had become *the* upper stage (the name was changed to 'Inertial' – characterizing the guidance system – to keep the same initials, IUS). And the supposedly cheap, temporary vehicle started getting expensive.

The original cost estimate in 1978 was $285 million. By 1980, the Air Force had renegotiated the contract price with Boeing to $439 million, with a 'ceiling' of $462 million.

In 1981 the 'ceiling' was raised to $518 million. What's more, the Air Force decided it had better hedge against further delays in the programming by buying $109 million worth of backup upper rocket stages from Martin Marietta (the first IUSs are supposed to ride on the top of the Air Force's Titan 34D rocket; the Martin Marietta upper stage could substitute on those missions, but not on space shuttle missions). The most frequent explanation for the cost overruns is that a project that started out as a simple, low-risk job ended up 'pushing the state of the art.'

Situated on a flat valley floor south of Seattle, the Boeing Space Center is a complex of two-story blocks painted in what I have come to think of as 'Aerospace Tan' – a color that seems to be popular with air-and-space factories up and down the West Coast. The complex also sports another sign of today's aerospace plant – the ubiquitous white satellite communications dish.

'Space Center' is something of a misnomer, since one of the big projects at the plant is the Air Force's air-launched cruise missile, which doesn't go into space at all. But, as Boeing public affairs officer William Rice explained to me, 'Much of Boeing's work is co-located in the Puget Sound area [aircraft plants in Everett and Renton, for example], which is important when you compare it with some other firms that are spread around the country. We have engineers coming off our commercial aircraft projects right on to defense projects, like the cruise missile. It's a reservoir of manpower. The same with the subcontractors who have learned to work with us over the years on airplanes – the TRWs, the Delcos, the Hamilton-Standards – are no strangers to us when we start working on an IUS.'

Rice and I walked through the building where engineers and technicians are testing and checking out the IUS. The IUS when assembled is a bell-shaped package more than 16 feet long and more than 9 feet in diameter. The package is made up of a large rocket motor shaped like a flattened sphere, with an exhaust cone at the base, a second, smaller rocket motor stacked on top of the first, a cylinder holding the two stages together, and, encircling the top of the package, a ring of instruments that is to support the satellite both literally and figuratively. That is, the 'equipment support section' is both the interface between the 'payload' satellite and the brains of the rocket ship.

In one room this equipment support section was wired to a bank of computers, whose screens were monitored by a white-coated technician. The ring of instruments contains batteries, an inertial guidance system (a complicated gyroscope, you might say), star sensors (like mariners of old, the IUS will navigate by the stars), space-to-earth communications systems, computers to tell the rocket motors when to ignite and when to separate, and very small rocket thrusters for steering. The computers were feeding it data simulating a real mission, to see whether all the electronic systems functioned as they should.

Rice pointed out that Boeing's contract with the Air Force required the IUS to be extraordinarily reliable: 'We have to make it ninety-six percent reliable,' he said, 'but we're at ninety-eight percent, which is unheard of for this type of spacecraft. Everything's redundant but the rocket motors themselves. The computers can analyze problems, turn over functions to alternate systems, or switch back and forth between systems for the best solutions. And the complete mission is built into its computers; once it's launched, it's autonomous.' One of the problems costing a great deal of money and time has in fact been the complexity of the programs, or 'software,' to run those computers; the Boeing subcontractor for the software, TRW, has found it extremely difficult to keep the program package compact enough to fit into the flight computer.

Rice pointed out another test room, the anechoic chamber. The walls and ceiling of an anechoic chamber are completely lined with thousands of pyramids of a special carbon material that doesn't reflect electromagnetic waves; it's something like a soundproof room, but the 'noise' in this case is in the form of radio waves. Here another equipment support package – ring of instruments – was being subjected to the electromagnetic environment that it would experience on a rocket launch pad. Tests would determine whether the external radiation would foul up the delicate electronics of the vehicle.

In another large hall sat a giant steel sphere, a vacuum chamber in which technicians would subject the IUS to the airlessness and heat and cold of outer space. In yet another room we donned white plastic coats to protect an assembled spacecraft from our lint: hard-hatted technicians, suspending the package from a crane, were balance-testing the craft before sending it off to Cape Canaveral.

Unlike previous rocket booster upper stages, the IUS burns solid, not liquid, fuel. Boeing persuaded the Air Force that solid-fueled rocket engines would be simpler, more reliable, cheaper, and safer than liquid-fueled ones. Unfortunately, the rocket motor subcontractor, United Technologies' Chemical Systems Division, found that the motor burst in a high-pressure test and that the solid fuel tended to develop cracks after it was inside the motor casing for a while. (Cracked solid fuel can cause such uneven burning as to throw the rocket off course or even explode the motor.) More cost and delay.

NASA had originally planned to buy a special package of three small IUS motors to carry out its future deep-space rockets, but in 1981 it decided instead to adapt an older upper rocket stage to its purposes – the liquid-fueled Centaur. This will further drive up the cost of the IUS to the Air Force.

NASA's delays in the shuttle program can also be blamed for some of the IUS problems, since the IUS had to be designed for a vehicle – the shuttle – that didn't really exist yet. Over three years something like 150 changes had to be made to accommodate changes in the shuttle. For example, according to earlier plans, astronauts were supposed to separate the IUS from the shuttle cargo bay by manipulating it with the remote-control arm in the shuttle. The latest plan is to eject the IUS with a spring, in order to get it away from the shuttle before its rocket motors ignite. This change added weight and design costs to the system.

According to the Air Force, problems in pushing the frontiers of space technology are not the only ones that delayed the IUS project. Rice said, 'The huge cost increase is because of the challenging we've been doing to the state of the art . . . it's not been an easy development. We did not realize the problems we would face in meeting the stringent standards the Air Force is imposing on the system. Boeing freely admits it was overoptimistic. It really was.'

The Air Force tells it slightly differently. In the spring of 1980, Brigadier General Spence M. Armstrong, from the R-and-D offices of Air Force headquarters at the Pentagon, went before the House Armed Services Committee to explain why the cost of the IUS was going up another $39 million. He outlined all the technical problems, but there was more. In the first place, he said that when the 'Interim' Upper Stage became the 'Inertial' Upper Stage, the Air Force thought it had only twenty-seven months left to build and launch the first two vehicles. 'Consequently,' he admitted, 'we developed a success-oriented program schedule which probably contributed to cost projections that were unrealistically low.' Another case of 'buying in'?

General Armstrong pointed out that the contract had been structured so as to encourage Boeing to put performance before economy. If Boeing met the target cost, it got a 7 percent fee on the contract. If it ran over the target, it paid for only 10 percent of the overrun, the Air Force picking up the other 90. And Boeing was liable for that 10 percent only up to the point where its service fee was reduced to 3 percent, that is, it was guaranteed a profit of at least 3 percent of the total contract, no matter how far over cost it ran. On top of that, Boeing would get *another* 7 percent if it met or exceeded the performance goals for the IUS.

So, Armstrong went on, Boeing had acted as one might expect: it spent more money so as to meet the performance goals and win the extra fees. At first, the Air Force thought it could contain the cost overruns. What they hadn't noticed was the degree to which Boeing had bought in on the project. Having accepted a contract at a price below its own cost estimates, the Boeing management put too few people on the job, which only increased the long-run problem. Armstrong said that part of this corner-cutting included failure to use full-time managers to watch the subcontractors. The Boeing managers, thin on manpower, 'also performed a minimum amount of internal and schedule planning, tracking, and analysis.'

The upshot of all this was that the Air Force insisted on renegotiating the contract with Boeing, raising Boeing's share of future overruns to 30 percent and limiting the government's ultimate liability, although those 'ceilings' seem to keep rising. According to General Armstrong, the result of the negotiation was that Boeing appointed a new IUS program manager and started performing all those management tasks it had neglected before.

On the other hand, according to Boeing Aerospace vice president Mark K. Miller, 'We had competent management from the beginning.' The changes in management were done 'to fit the changing phase of the program . . . you can't relate it to the problems.' Take your pick.

The Inertial Upper Stage is certainly not the only Boeing interest in the U.S. space program, nor the only one related in some way to the space shuttle. Boeing Services International has a contract with NASA to provide ground support, supply, and transportation for the shuttle. About 1,900 Boeing employees manage the inventory of shuttle parts and see to their shipment. At Kennedy Space Center, the NASA shuttle launch site, about 400 Boeing employees 'supply nearly everything . . . except the alligators,' as a Boeing press release puts it.

Boeing has also had contracts with NASA to study the successor to the IUS – the genuine, reusable space tug that NASA originally wanted for the shuttle. Without such a rocket, it won't be possible to use the shuttle in its vaunted role of servicing and repairing satellites at higher altitudes. Nor will it be possible to build space stations at the high, earth-stationary orbit, 22,250 miles above the equator.

Ultimately, the effective use of the space tug, or 'orbital transfer vehicle,' would require the building of yet another NASA-Boeing project: the Space Operations Center. The Space Operations Center (SOC) would be our first permanent space station, about 200 to 250 miles up. The shuttle would carry modules – service modules, supply modules, living modules – into orbit, where the crew would assemble them into a space station. The eight-person crew could service and launch space tugs, refuel or service satellites, or carry out experiments. Eventually, an enlarged space station would become the construction base for even bigger projects – large radar or communications antennae, or even space stations in the high, stationary orbit. In 1981, Boeing foresaw two or more contractor teams sharing $20 to $30 million for competitive design work on the Space Operations Center. Boeing analysts think that the space station could be in orbit by 1989 for anywhere from $3 to $7 billion, depending on how ambitious the design was.

Parts of Rockwell International Corporation go back a long way in the space business. North American Aviation built the Apollo moonshot command ships. North American's Rocketdyne Division built the giant F-1 rocket engines for the Saturn V rockets that boosted the moonships into space. In 1967 North American Aviation and Rockwell Standard merged to become North American Rockwell. Reflecting the conglomerate nature of the new firm, in 1973 it renamed itself the Rockwell International corporation. By the 1980s Rockwell was a $7-billion-a-year multinational, with five operations sectors made up of some thirty divisions. The North American Space Operations group has three divisions: Space Operations and Satellite Systems, Rocketdyne, and Space Transportation System Development and Production.

The Space Transportation System Development and Production Division has the prime contract for the space shuttle orbiters; Rocketdyne makes the powerful triplet of main engines on each orbiter. In fiscal year 1981, the shuttle made Rockwell NASA's biggest prime contractor by a factor of more than five. NASA awarded Rockwell $1.47 billion that year (the runner-up was Martin Marietta, which makes the shuttle's huge external fuel tank). In the same year, Rockwell had Defense Department prime contracts worth $1.12 billion, making it the sixteenth-largest DoD prime contractor and the only one to sell more to NASA than to DoD.

More than 200 companies have some stake as subcontractors in the space shuttle. Two thirds of the members of the Aerospace Industries Association, the chief aerospace trade group, hold shuttle contracts. But with revenues at roughly $1.25 billion a year, Rockwell obviously has the greatest single stake in the future of the shuttle, and therefore in the future of the U.S. space program. That commitment may have something to do with the relative openness of Rockwell officials in advocating their views of where we ought to be going in space.

All the major aerospace companies maintain Washington lobbying operations. Moreover, to compete for Defense Department contracts, you don't sit and wait for the Air Force to come to your door asking for a specific piece of equipment. You help them define what they'll 'need' in the first place – identifying what's technologically possible, what it can do for them, what it might cost. Since the development of new space (and other defense) equipment takes five to ten years, an aerospace company *must* try to anticipate – and shape – future policy choices. The only difference among companies is whether or not they expose their policy preferences in the public record.

Rockwell is refreshingly frank about advancing its vision of our future, civilian and military, in space. In 1981 Rockwell Space Operations and Satellite Systems head William Strathern told a visiting Congressional committee, 'Rockwell, obviously, has been spending some time over the years thinking about – hopefully on a bipartisan basis – what the U.S. space program should be.'

At Rockwell's Downey, California, plant I heard some of those thoughts about the future of the space program. As prime contractor for the space shuttle orbiter and main rocket engines, Rockwell puts all the parts together at its Palmdale plant (a factory space leased from the Air Force), not far from Edwards Air Force Base. Rockwell has farmed out various pieces of the orbiter to subcontractors, but it does manufacture the forward compartment, or crew module.

I spoke first with William Rhote of the Space Transportation System Development and Production Division. Rhote is a middle-aged, chain-smoking man with a youthful enthusiasm for the technology he works with. He conducted me through the vast factory floor (about four football fields of open area). We walked along a carpeted path bisecting the work areas. In the brightly lit distance I made out the skeletal form of a shuttle orbiter nose. At dozens of work stations craftsmen and technicans machined parts, drilled, fitted, huddled over computer video terminals. Making high-tech space hardware is perhaps the real handicraft of our age: the products require the work of thousands, coordinated by corporate discipline, yet everything seems to be made by hand, crafted with infinite care. (One of these craftsmen had a stenciled sign hung out on his bench: YOU CAN'T SAY CIVILIZATION ISN'T ADVANCING: IN EVERY NEW WAR THEY KILL YOU IN A NEW WAY.)

Past the fabrication area was the display-and-lecture hall. Placards and models in window displays explained the shuttle and other Rockwell projects. The overwhelming centerpiece of the room, however, was a full-sized model of the shuttle orbiter body. We climbed into the crew quarters, directly under the cockpit. It seemed a very small room to house six or seven people, but Rhote remarked on how spacious it was. Relatively speaking, he was right: at the other end of the hall was an Apollo moonshot capsule, and it was dwarfed by the orbiter. More impressive for its spaciousness is the orbiter's cargo compartment, about the size of a boxcar. It's that cavity that has stirred the imagination of space enthusiasts, civilian and military.

Rhote spoke of the opportunities he and others at Rockwell think the shuttle presents, especially for the military. Like Boeing, Rockwell has been studying the Space Operations Center – the low-orbit, eight-man space station – for NASA. Rhote thought that this space station would be the next major national space goal and that the Reagan Administration would soon announce it as such (in fact, Reagan's fiscal 1983 budget allowed NASA to go on studying the idea, but committed little money to it). Separately and together, the shuttle and the space station offer real promise to the military, Rhote thought.

'If you want to build large antenna structures for the military [say, for a radar scanner aimed at the earth], the Space Operations Center is where you do it from. There you build and check out the structure, then you put it on an orbital transfer vehicle to place it in a high-inclination orbit.' (High-inclination orbits, passing near the poles, allow coverage of the whole world.) The low-orbit space station would also be the staging point for building space stations in the high-altitude, earth-synchronous orbits.

Rhote said, 'The geosynchronous slots are getting more and more crowded. Why not have one big vehicle as a communications platform, a kind of communications factory? And you'd need a crew to keep it functioning. The Air Force could use the maintenance facility in space – or the ability to bring the high-altitude satellites back down for work.' He thought that the first U.S. geosynchronous platform would be a service station for both military and civilian satellites – a fact that we might or might not tell the world: 'To the Soviets we're doing strictly scientific and commercial work – there's no way they can dispute us on that.'

Rhote expects the shuttle to benefit the military long before space stations become a reality: 'Not only could the Air Force refurbish their satellites [by resupplying them in space or bringing them down for bigger jobs] but the shuttle offers other new possibilities. Weight in a satellite is no longer a problem. They could make a satellite with cast iron and vacuum tubes if they wanted to. They can design it to be brought back for refurbishment. The cost of payload design will go way down . . . the Air Force is already designing satellites to shuttle parameters.'

He foresees a distinctively *military* space shuttle for the future. 'The military,' he said, 'would like to have two things from the shuttle: first, quick-reaction capability – the ability to take off on short notice; second, survivability – the ability to resist attack, including nuclear attack. The communications will be more secure against electronic countermeasures. It will have more delta-vee capability – more maneuverability, lighter. It'll have quicker turnaround. The military are willing to take more chances than NASA is to get a quick-reaction capability. The military shuttle will fly higher-inclination orbits. It will be autonomous: it won't be shoved around by Houston control or

CSOC – the new Consolidated Space Operations Center; it'll be less dependent on ground stations.'

While Rockwell is working with Air Force Space Division on future shuttle ideas, Rhote was disappointed that the Air Force was not moving faster. 'The go-go attitude of the space cadets isn't the way to go, but neither is the conservatism of some of the military types. The answer lies in between. One of the key drivers in reaching that middle ground is the conclusion that there's got to be a separate Space Command. The military have to have their own shuttle fleet, to do things their way.'

But the military, said Rhote, have been slow to push for all this. He said, 'It's frustrating, as the country's leading manufacturer of spacecraft capabilities, we see how things should be for the best defense of the nation. We work the problem continuously. It seems like we sometimes lobby more than we market. We'd rather they came to us with what they want. But the Pentagon moves slowly . . . it's a problem of large bureaucracy . . . in the Pentagon there is a large body of enlightened, forward-looking people, and they're frustrated too.'

Rhote nonetheless remains optimistic. 'Things are going to change in the next ten years. The American public understands the potential of space better than some people in the Air Force. We have former astronauts in the U.S. Senate . . . their eyes light up when you say that the shuttle is a military vehicle. As the shuttle improves in the next few years, military enthusiasm for it will grow. Polls we've commissioned show that people recognize the importance of space for the military. Congress is pushing the military use of space because the public is for it.'

Rhote is confident that 'it's only a matter of time. We're anticipating that, defining what the military orbiter will be like. We see three military shuttle orbiters before 1990 – we call them Orbiters 107, 108, and 109.'

If there is a truer believer in the exploitation of space than Bill Rhote, it is 'Sandy' Sanborn. Colonel Morgan W. Sanborn, USAF (ret.), is Manager of New Business Requirements for Rockwell's Space Transportation System Development and Production Division. Sanborn's physique makes for an imposing presence: he is a tall man whose dark eyebrows curl up at the ends, giving him an almost theatrical appearance. His calendar for the month showed a number of speaking engagements. I guessed that his spontaneous excitement about his subject made him a popular public speaker, and one Rockwell was happy to send out to spread the word about the potential of space.

Sanborn's experience in the Air Force has led him to agree with Rhote's concerns about the lack of progressive thinking about space in Air Force upper echelons. He recalled, 'While I was last assigned to the Pentagon, I was chairman of a mission evaluation group – Carter was trying the so-called zero-based budgeting in which you are supposed to re-examine every program from the ground up. In arguing for space capabilities, I found a comparative lack of support and understanding for what you can do there. . . . We've had some increases in the budget, yes, but I'm tired of seeing a small cadre of officers having to fall on their swords again and again to get the right thing done. There are individuals with advanced ideas in the Air Force, but if they scream too loud they don't get promoted. . . . The operational commands have the manpower to develop and document their policy recommendations. But the users of space systems weren't out there helping us.'

Sanborn thought that people in industry were ahead of the Air Force in thinking about the future of space. 'In industry,' he said, 'people who have been in the business a long time have seen the evolution – how systems that almost got killed became crucially important – are convinced, whereas military people rotate in and out. There's more capability in industry to appreciate the longer-term aspects, the philosophy. And it's not just financial motivation – sure, everyone wants to get promoted, to see his programs succeed – but people are also equally motivated to do what they think is right.'

Sanborn then warmed up on the subject of the military benefits from the use of space, and especially to the potential of the shuttle era. 'You know, a commander has to know where the enemy is, where his own troops are, and to communicate with his troops. In the past, it's always been essential for military systems to be *fast* – like a football back taking advantage of a quick opening in the line, to be *accurate*, and to have *range*. In the past, these qualities have been limited to a local theater – although the airplane broadened that somewhat, but we need as much of these things as we can get, on a global scale. We have national interests everywhere on earth; we may need to react anywhere.

'For example, beyond the shuttle we need an on-demand launch vehicle, to get anywhere on earth within ninety minutes. It needs to be fast, accurate, global. I do briefings to military people on this and get no response at all. They always want to see the scenario.' He thinks that any specific scenario for using new weapons is going to sound implausible, but that as a rule military surprise produces *faits accomplis* – for example, when the Soviets moved into Hungary in 1956 and into Czechoslovakia in 1968. 'With space reconnaissance and surveillance, with rapid response, maybe you can stop or deter them.'

One of Rockwell's major projects is the Global Positioning System, a network of satellites that will permit unprecedentedly accurate navigation. 'With GPS,' says Sanborn, 'we'll be able to launch conventional [nonnuclear] weapons from orbit with pin-point accuracy. I remember flying F-4s in Vietnam – they rarely hit their targets, it was really tough. With a space-based missile launch – using the GPS and conventional weapons – you can signal your intent but withhold commitment . . . you've got flexibility. It's not *only* your capability; more important is the other guy's *perception* of what you might do. There is psychology in all this. Your advantage or lack of it depends on what the other guy's got. If he has a space-based weapon that gets there in sixty minutes, and yours gets there an hour later, it's too late.'

Underlying Sanborn's arguments about the advantages of space weaponry is a fundamental belief in a need for the United States to seek a military advantage over the Soviet Union. 'We need the military to help preserve the nation, to protect our vital interests. If we didn't have the military we do have, the whole Middle East would already have been sucked up by the Soviets. There's that Brezhnev quote – their long-term plan is to control the treasure chests of the West. . . . When people ask how much is enough for military spending, they compare it to welfare or something else we're doing. I can't understand that. The question is not how much do we spend in comparison to education and welfare, but how much in comparison to what the adversary spends.'

Sanborn has a standard briefing he has given to military people to the potential of the space shuttle and later versions of it. 'Most people in the Air Force aren't eager to have the Space Transportation System at Vandenberg – they'd just as soon stick to the Titan [expendable rocket booster]. But I try to explain that the United States could win in a technological competition with the Soviets if we were committed. The evolution of military capabilities is partly not under our control: we have to do things because the Soviets are. But if we set our minds to it we could get a decisive advantage.'

Sanborn has worked to develop a Rockwell proposal for the future of U.S. national space policy, civilian and military. He frames the importance of his proposal in large terms. He pulled from his files a copy of J. B. Sparks's 'Histomap of History' – a Rand McNally chart that graphically depicts, among others things, the expansion and decline of major civilizations. Pointing to the bottom end of the chart, he said, 'Past civilizations have risen and fallen and the west seems to be in decline. The U.S. needs to do something to reduce this decline and the ascendancy of the Soviet Union – a bad trend for our nation. Space is an area where we might establish new goals, galvanize public opinion, regain our momentum. It can attract public attention, excite them.'

The Rockwell space-policy plan, part of which Rockwell official William Strathern had presented to the House of Representatives space subcommittee, is an ambitious one indeed. The plan is explained in a Rockwell booklet, 'Space: America's Frontier for Growth, Leadership and Freedom,' in which Sanborn had a major hand. The booklet sees the United States faced with several unfavorable trends, which will lead the country '. . . to find itself in an increasingly precarious position, beset with problems both at home and abroad.' These trends, it says, include decline in economic growth, growing dependence on imported fossil fuels, loss of military advantage, and decline in national morale. Space technology, says Rockwell, can make a major, perhaps decisive, contribution to the reversal of these trends.

Operating 'at the very frontier of technology,' the space program 'spawns technological advancements that ultimately help to foster higher productivity, open new markets, and develop new products.' Strathern appealed to the Congressmen for a government policy to encourage the industrialization of space: more rapid depreciation, tax credits for research and development, a tax moratorium for initial space products and services, expanded laws that allow private enterprise in space, and . . . 'a space plan that the government sponsors in conjunction with industry, so that the industrialists understand over a period of time what they can expect to happen in space.'

Space systems, says the Rockwell booklet, can also provide some solutions to our energy problems. Satellite reconnaissance can be used to discover and monitor natural resources. And 'space development also has been a technology driver which promises new techniques and materials which will apply to our energy production, distribution and conservation efforts.' Finally, in the longer run, solar energy from space – gathered and transmitted by enormous solar-power satellites – 'may be able to support a significant part of our demand, that of our allies, and of the less-developed nations.'

As to defense, the Rockwell people see space systems potentially giving the United States military superiority over the Soviet Union. As with its proposals for the economic exploitation of space, Rockwell suggests that this superiority can be developed in three stages over the next thirty years. In the 1980s, the United States continues to improve on what the military call 'force multipliers' in space: using satellites for reconnaissance, navigation, weather forecasting, and communications in ways to improve the effectiveness of the land, sea, and air forces. In the 1990s, the proposal goes on, we can progress so far that space systems offer 'decisive support' for our military forces, while establishing the superiority of the U.S. military in space. By the end of the first decade in the twenty-first century, we can be in the stage of 'hostility containment.' 'In summary,' Rockwell says, 'military space systems offer one credible way to counter the preponderant USSR force. A technological leapfrog on our part could render USSR conventional forces obsolete and once again place the free world forces in a decisive position.'

The Rockwell plan holds that three particular space programs will provide the keys to the American economic and military renascence. In the eighties, the plan calls for the full development of the shuttle and of related rockets for getting big payloads into space. Rockwell has several proposals for new rocket boosters – among them are improving the space shuttle main engines for greater lifting ability, building an unmanned rocket using the shuttle rocket engines to carry tremendous cargo loads into orbit, and building the orbital transfer vehicle, the manned 'space tug' that NASA had originally envisioned for getting satellites into high orbits.

This space tug, or OTV, is most important, because it links the other two main projects in the Rockwell plan. Rockwell proposes that in the 1980s the U.S. build a low-earth-orbit space station. What Boeing called the Space Operations Center is called the Earth Support Base by Rockwell. Rockwell division chief Strathern told the visiting House subcommittee: 'If you built a platform that you could use for defense, you could also use it for industry and science. Three units would be coming together, as opposed to trying to build three different ones for three different groups.'

The third large project, materializing in the twenty-first century, would be a geostationary space base – a large fortress 22,250 miles up hovering at a point over the equator. Such a space station, says Rockwell, promises 'Global Battle Management capability. . . . Comprehensive multispectral surveillance of worldwide military activity would virtually preclude military, technical, and political surprises in the future.' But the station is more than an all-seeing watchtower: the space station will be the ultimate military command post, '. . . permitting direct, rapid, and reliable command and control of all military forces.'

'The command post,' Strathern told the committee, 'will be protected by some kind of laser device.' The president of Rockwell's Rocketdyne Division, Norman Ryker, told the same committee: 'Rocketdyne is one of the two major high-energy-laser suppliers. We feel that there is rationale for placing a high-energy-laser demonstrator in space and that the United States is in a position to accomplish that if a national goal is established.' This 'demonstrator,' though, is only the beginning. The promise is that space-based lasers will defend not only our satellites but our territory, protecting us from ballistic missiles and airplanes. When we have reached that stage of invulnerability, we will be in the period of 'hostility containment.' The result, one of the Rockwell briefing charts suggests, is that we have 'full control of U.S. destiny' and 'long-term security for earth's population.'

And all of this, it turns out, will be relatively cheap. Rockwell recommends committing one third of 1 percent of our Gross National Product to space development, civilian and military, over the next thirty years. This is in fact less than we are now spending, even if you don't count the defense programs whose budgets are secret. (The fiscal year 1983 NASA and Defense Department space budgets added up to more than $15 billion, which would be three to four tenths of 1 percent of 1981 GNP.) Given that some estimates place the development of a complete solar-power satellite system at upward of $1 trillion and that an anti-ballistic missile laser system might cost $200 billion or more, it's hard to see how the Rockwell ambitions can be supported at funding levels that rise to only $26 billion by the year 2020.

Setting out to use space technology to solve our economic and military problems, says the Rockwell booklet, will give America a new sense of national purpose. 'We must strive toward a world where people can live in harmony and sufficiency, and where their emotional needs can be satisfied. A coherent national space program has in the past helped provide for these social needs, by providing technology and economic drivers, by establishing concrete, attainable national goals, and by maintaining for the U.S. a place of pride and leadership . . . in the future, [the uses of space] . . . can offer avenues for solving our social problems.'

# CHAPTER THREE

# The C-Cubed Revolution: Space Communications

In *War and Peace* Tolstoy pictured Napoleon's command of the Grande Armée against the Russians at the Battle of Borodino, September 7, 1812. He described the orders Napoleon laid down before the battle began:

> . . . the disposition stated that *after the action has begun on these lines subsequent commands will be issued in accordance with the enemy's movements*, and therefore it might be inferred that Napoleon would take all necessary measures during the progress of the engagement. But this was not, and could not be the case because during the whole battle Napoleon was so far from the scene of action that (as it appeared later) the course of the battle could not have been known to him and not a single instruction given by him during the fight could be executed . . .

> From the battlefield adjutants he had sent out and orderlies from his marshals were continually galloping up to Napoleon with reports of the progress of the action; but all these reports were deceptive, both because in the heat of the fray it was impossible to say what was happening at any given moment, and because many of the adjutants did not go to the actual place of conflict but simply repeated what they had heard from others; and also because while an adjutant was riding the couple of miles to Napoleon circumstances changed and the news he brought was already ceasing to be accurate. Thus an adjutant came galloping up from Murat with tidings that Borodino had been occupied and the bridge over the Kalocha was in the hands of the French. The adjutant asked whether Napoleon wished the troops to

cross the bridge. Napoleon gave orders for the troops to form up on the farther side and wait. But before that command was given – almost as soon in fact as the adjutant had left Borodino – the bridge had been retaken by the Russians and burnt . . .

Another adjutant rushed up from the fleches with a pale and frightened face and reported to Napoleon that their attack had been repulsed, Campan wounded and Davout killed; while in fact the entrenchments had been recaptured by other French troops (at the very time when the adjutant was told that the French had been driven back), and Davout was alive and well except for slight bruising. On the basis of such inevitably untrustworthy reports Napoleon gave his orders, which had either been executed before he gave them, or else could not be, and never were, executed.

A modern military analyst, reading Tolstoy's fatalistic account of the battle, might be tempted to conclude that Napoleon's main problem was a lack of good C-cubed.

'C-cubed,' or 'C³,' is the latest U.S. military jargon for 'command, control and communications.' Modern military commanders can communicate with their troops by means undreamed of by Napoleon or Tolstoy. Tolstoy's whole point in relating the Battle of Borodino was to show that military commanders are *not* in 'command and control' of their forces, that '. . . the way in which these men slaughtered one another was not decided by Napoleon's will but occurred independently of him, in accord with the will of the hundreds of thousands of individuals who took part in the common action.'

Since the advent of telephone, telegraph, and wireless radio, military communications have vastly improved over Napoleon's. The modern military commander has a much better chance to find out immediately (or, as the jargon has it, 'in real time') whether his orders have been carried out and what is happening at the moment.

Equipped with high-frequency radios and perhaps some field telephones, Napoleon could have received frequent reports from his officers at the scene of battle, and he could have revised his orders accordingly. The chances would have been increased that his subordinates did what they did on his order or with his knowledge and approval.

Even in the nineteenth century, wire telegraphy began to extend the powers of command and control of military officers. But wireless radio set military communications free of the preset structure of telegraph lines: by bouncing radio signals off the ionosphere (the electrically charged upper layers of the atmosphere), military commanders could keep in touch with the troops even over long distances. And using combinations of land lines, undersea cables, and numerous radio relay stations, it even became possible to encircle the globe with military communications networks. Thanks to the worldwide commitments and military bases it ended up with after World War II, the United States built the most extensive network of all.

In the 1960s, that network still relied mostly on undersea cables and radio relay stations. In 1968 North Korean gunboats attacked and captured an American electronic spy ship, the U.S.S. *Pueblo*. The captain of the ship radioed a call for help. His message went first to the headquarters of the Commander, Naval Forces Japan. Despite a 'Pinnacle' designation, meaning that it was supposed to go directly to Washington, the message took fifty-five minutes to 'process' before it went out on the Defense Communications System lines to the Pentagon. There, another hour and sixteen minutes passed before the Joint Chiefs saw the message (when it got to the Pentagon, it bore the 'Flash' designation – it should have gone upstairs at once). Had the President wished to order immediate military action to aid the *Pueblo*, it would have been much too late.

Ten years later, Gerald Dineen, the chief Pentagon civilian for communications, contrasted the *Pueblo* incident with the *Mayagüez* operation. The *Mayagüez* was a U.S. merchant ship seized by the Cambodians in May 1975. President Ford decided to order a military rescue operation. The rescue succeeded, but because of a helicopter crash, 38 American soldiers were lost for the 39 *Mayagüez*

crewmen saved. (And there were some reports that negotiations were about to succeed in getting them released
without military action.) Dineen pointed out in 1978 that
the Pentagon was '. . . able to communicate . . . from the
National Military Command Center using a NATO satellite and our own DSCS II satellite – Defense Satellite
Communications System II – using secure voice con-
ferencing. We were able to get the Chairman of the Joint
Chiefs and the other Chiefs to communicate by that
method with the other people on the scene and conduct
that operation successfully.'

In the past few years the United States has relied more
and more on satellites to carry its military communications
around the world. In fact, about two thirds of the Defense
Department's long-distance communications go by satellite. Joseph Toma, head of the C3 Systems Evaluation
Office, under the Joint Chiefs of Staff, said, 'The shift to
satellite communications systems has happened for a lot of
good reasons. The costs of what you can get through the
satellites versus putting in new cables, for example, have
gone down greatly. The key factors are quality and capability. Satellite systems are more flexible. It's true that an HF
[high frequency] radio is very flexible, but is subject to all
the vagaries of the ionosphere, and time of day, and time of
year, and what the sun is doing, and has relatively low data
rates. So there's been a tendency to go to satellite systems,
not just because of their quality and capacity for transcon-
tinental purposes on a fixed-station basis, but because of
the flexibility of the mobility of the equipment . . . you can
take along a dish antenna a few feet in diameter, set it up
somewhere, and you're in business.'

There is also a kind of Parkinson's law in the communications business. As Toma remarks: 'As with all technology,
I guess, the more people use it, the more people want out of
it. When we were able to get high-quality data transmissions, for example, pictures out of Southeast Asia during
the latter stages of the Vietnam war, the decision-makers
here in Washington tended to want that information more.'

Another Pentagon official who works on 'C-cubed' is the
Assistant Deputy Under Secretary of Defense for Com-

munications, Command, and Control, Dr Thomas P. Quinn. Quinn looks on the growing demand for satellite communications services as a part of a larger trend in military history: 'The purpose of the C-cubed system is to allow the commander to exercise the forces. Since we started having wars, there has been some kind of sensor out there examining what's happening on the battlefield: the sensor communicates information to a command center of some kind, a decision is made, the decision is then communicated back out to the forces, then the forces do something in response to that decision. The sensors take a look at what has happened as a result of that, and the whole thing reiterates. In early times, the sensors were men on the field, the communications systems were runners. As we progressed – in World War One, for example, we got some optics that improved the sensors a little bit, motorized runners improved the communications – the scope of decision of the commanders was expanded. By World War Two, we had airborne sensors, radar; communications were improved by the telephone, some kinds of radio; the decision process was more formalized, bringing larger numbers of commanders into the picture. The scope of control was expanded.

'Once, the field commander might be the only one involved in decision-making for many days, or even months at a time. Today, we do the sensing with satellite systems as well as radars. We communicate with satellite systems, radio systems; our decision-making process is greatly enhanced with computers, automated display systems, and so on. Now the decision involves in some cases the highest authorities in the country – in the case of nuclear war it would involve the President himself. So what has happened historically is that the scope of battle being examined was very narrow at the beginning – the human observer might look at a couple of hundred yards. Later, the guy with good binoculars could go half a mile. In World War Two, we're still talking about distances of a hundred miles or so. But the trend is that the amount of information going into the process has been expanding as the scope that one can see and the span of control expands. Today, the decision-makers are dealing with information not only from the sensors, but from the intelligence community, from political analysts.

'That drives you into finding newer communications systems, better sensors. Satellites are a natural way to handle very large volumes of traffic in moving information from one place to another. So satellites have naturally evolved to fill the need created by the other developments of warfare and command and control.'

Toma, in the Joint Chiefs' office, said that one practical example of the kind of satellite use Quinn was talking about '. . . would be our use of C-130 aircraft as airborn command posts when we hold our annual Reforger exercises in Europe. The aircraft might be directing the fighters under its command with line-of-sight high-frequency radio, but it uses a satellite link to keep in touch with its airbase or even, in some cases, CONUS [Continental U.S.].'

He also gave examples of real operations in which the Pentagon used 'mobile' or 'deployable' satellite terminals. 'A couple of years ago we had a hurricane in the Caribbean that damaged the Dominican Republic and several other islands rather severely. We ran some emergency airlift operations down there. Since there are very poor local communications on that chain of islands, we put about six satellite terminals there to coordinate our airlift.

'Back when there was a rebellion in Zaire we thought we might have to evacuate some of our citizens; we didn't know if that would be a shooting situation or not. As it turned out, we airlifted supplies in to help the French help the Zairean government. We deployed some satellite terminals at several locations in Africa so we could put our air controllers in.

'After the Jonestown massacre in Guyana, the U.S. Air Force had to go down and fly out a lot of the bodies. Since communications were relatively poor down in that area, we sent a couple of satellite terminals with our airplanes so they could communicate back to Panama, where the Southern Command Headquarters were. And the Department of State was able to use those terminals too.'

Toma said that the military would similarly move portable terminals into a battle situation, especially where an established communications network didn't exist or where the network had been destroyed. 'You remember the evacuation of Saigon: at the end, our underwater cables were cut, and the last communications link was by satellite, both between Saigon and back here and between Saigon and the ships offshore that they were coordinating the evacuation with.'

The use of satellite communications varies among and within the services. All the major U.S. military command headquarters around the world – about twenty-six of them – are linked in the WWMCCS (sometimes pronounced 'Wimex,' the Worldwide Military Command and Control System). That system includes not just commanders talking to each other, but computers passing around the great masses of data that make a modern bureaucracy run. The Air Force can talk to its airborne command posts by satellite, and in the past three or four years it's been equipping its strategic bombers, its tankers, and its missile sites with satellite terminals. Every major Navy ship has at least a satellite communications receiving antenna, and many can talk two-way as well. The Army has the farthest to go in equipping its field forces with portable terminals, but has big plans in that direction. Quinn estimates that by 1990 the U.S. military will have about 5,000 mobile ultra-high-frequency satellite terminals all together.

Suppose you had a piece of string about 2 feet long with a light weight, say a spool of thread, on one end. Then imagine holding on to the loose end of the string with one hand, grabbing the string a few inches from the spool with the one hand, and twirling the spool, just fast enough to keep it going around in a circle. Now, keeping the spool just barely in orbit around your hand, if you were to let out a few inches of string, you'd notice two things. First, the spool would slow down as the circle it travels in gets bigger. Second, you'd have to jerk the string harder – apply more energy – to keep the spool swinging around in a circle at all. Now try it in the other direction: with the hand on the free end of the string, start pulling the string through the other hand, the one with the spool circling it. You'd have to tug fairly hard – apply more energy – to pull in the circle's string-radius. And as the twirling spool got closer to the pivot hand, it would start moving around faster and faster. Think of the spool as a satellite, your hand as the center of the earth. The string was the pull of gravity (although we cheated and used it as a rocket too).

At any distance from the center of the earth that a satellite may orbit, there is a fixed *period*, or time, for one swing around. The closer to earth the satellite is, the faster it has to go around to stay in orbit. The farther away it is, the slower it moves. As it happens, a satellite going around the earth at an altitude of about 100 to 200 miles has a period of about an hour and a half. But a satellite orbiting at 22,250 miles up, like the ones whose weather pictures you see in the evening news, takes twenty-four hours – the amount of time it takes the earth to turn once on its axis.

If you could put a satellite into a circular orbit with a twenty-four-hour period, and the plane of that satellite's orbit was at right angles to the earth's axis (which is to say in a straight line from the center of the earth to the equator to the satellite), you'd have a *geostationary* orbit. The satellite's period would be synchronized with the earth's rotation. Although the satellite is zipping around the earth at a respectable 6,800 miles per hour, it's as though the earth were always turning to face it. In fact, it is. So from the earth, the satellite seems to be standing still over a point on the equator.*

In 1945 Arthur C. Clarke (scientist later to become science-fiction writer) published an article on extraterrestrial relays. He proposed a new radio communications system: three satellites spaced equally along the geostationary orbit. Each would carry a transponder, a combination receiver and transmitter, for passing signals along.

---

* A geostationary orbit is only possible along the equator. The plane of a satellite's orbit always passes through the gravitational center of the earth, and the only points on the earth's surface that also revolve around the earth's center are along the equator. Geo*synchronous* satellites in orbits that aren't in the same plane as the equator will appear to trace a figure-eight pattern above and below the equator. This is why it will never be possible to make a spy satellite hover directly over Moscow or Washington.

From almost any place in the world you could send a message to the satellite, and that satellite would retransmit your message to another ground station. You could bounce messages up and down all the way around the world. Or you might even send a message up, have the satellite lob it over to one of the others, then have that one send the message back down: point-to-point communications without a single ground relay station in between. The idea of extraterrestrial relays was a great one whose time came in less than twenty years.

The Pentagon started working on the Clarke idea in 1958. By that time, Bell labs and others had given a lot more thought to satellite communications, and the geostationary scheme was one of several. In 1946 the Army had gotten a radar signal to the moon and back, and by 1954 the Navy was bouncing voice messages off the same satellite. At the right times of the month, Navy researchers could talk to each other, via the moon, between Washington and San Diego, later between Washington and Hawaii.

The first artificial communications satellite was a one-way transmitter – from space to ground. It was President Eisenhower's taped voice wishing us all a merry Christmas. (The experiment was part of Project SCORE – Signal Communication by Orbiting Relay Equipment.) Two years later, NASA put up its own little moon – a 100-foot-diameter silver moon, floating 1,000 miles overhead. At least for a few minutes at a time, earth stations could bounce radio signals off it. The Air Force had a different idea: instead of a balloon, how about 500 million tiny copper wires in orbit? Project West Ford, they called it. Astronomers were aghast: what would that cloud of metal do to their view of the universe? After the first shot failed to get into orbit, the next finally made it in 1963. Both the Air Force and the sky watchers were wrong: the copper cloud drifted apart, making for neither good communication nor bad astronomy.

The Air Force and NASA experimented with other kinds of satellites that could actually receive and retransmit messages, but these flew at relatively low altitudes. The best long-run bet would be the geostationary orbit. In 1960 three separate military communications programs were combined into one project under Army management: Project ADVENT. The ultimate ADVENT satellite would weigh 1,250 pounds; rockets to put that much weight into geosynchronous orbit were still on the drawing boards at the time, and in fact came in later than planned. The cost of the satellite would be $352 million. The schedule slipped. Newer, transistorized designs could do a better job and would weigh less than half as much. Secretary of Defense McNamara killed ADVENT in 1962.

NASA came in first. It got the first geostationary satellite, SYNCOM 3, into orbit (two earlier attempts failed) in August 1964.

In those early years, it wasn't easy to put up a geosynchronous satellite. Getting the launch rockets to burn just right and the guidance system to perform perfectly is hard enough. You are shooting for a narrow lane in space, 22,250 miles away.

The usual technique is to use what is called a 'transfer orbit': the satellite is first launched into a highly elliptical orbit, ranging in altitude from a couple of hundred miles out to nearly 22,250. The transfer orbit isn't parallel to the equator, but more likely at an inclination of 25 or 30 degrees. This orbit takes best advantage of the extra toss given by the earth's rotation speed at Cape Canaveral. Later, after several orbits, perhaps, comes the 'apogee kick.' The farther a satellite is from the earth, the less energy it takes to change the plane of its orbit. The effect is something like pushing a child's swing: the best time to give it another shove is just as it's descending from the highest point of its arc, where it is moving slowly – its energy is stored, or 'potential' (gravity will accelerate it again), rather than moving, or 'kinetic.'

A geosynchronous satellite is equipped with an 'apogee kick motor' that will give it a small, precisely aimed shove at the apogee, or high point, of its orbit. If the rocket burns at just the right rate for just the right amount of time in just the right direction, the satellite's orbit is circularized (its apogee and periree, highest and lowest points, become virtually identical) at 22,250 miles and is made parallel with the equator. It becomes geostationary.

The satellite's controllers have to keep the satellite in the lane once they get it there. In the vacuum of space, it ought to keep moving in the direction you shove it, pulled into a circle or an ellipse by the earth's gravity. The earth, though, isn't a perfect, exact sphere. With the slight variations in gravity, it's hard to line the satellite up exactly with the equator and keep it that way. Then there's the moon's gravity to *perturb* the satellite's orbit, to distort the geometry of the space around the earth. And there's the solar wind, the invisible flow of subatomic particles streaming out from the sun. Over a long period of time, they too will alter a satellite's orbit.

Nor it is enough just to get the satellite up there: it's got radio antennae, and they have to be pointed in the right direction, toward earth, maybe even toward a specific spot on earth. Although the earth keeps turning to face the satellite, the satellite doesn't automatically keep turning to face it back. To point the satellite, you need small 'attitude-control thrusters.' And you need to stabilize it, keep it from rolling or wobbling. This usually calls for spinning something on the satellite like a gyroscope. In sophisticated designs, the antennae platform is 'despun' – it stays still while the rest of the satellite spins around it. Finally there's the radio equipment. It's difficult and expensive to get satellites that high, so the electronic gear has to be small and light. But if it's going to be worth the expense, it also has to be able to relay lots of messages at once. That's where transistors and, later, microcircuits come in.

The first operational Defense Department communications satellites worked around many of these problems. These were the satellites of the Initial Defense Satellite Communication Program (IDSCP). As early as 1961 the Air Force had The Aerospace Corporation looking at a system of 100-pound satellites that would orbit at only about a 5,000-mile altitude. When McNamara canceled the Army's ADVENT, the Air Force plan became the prime candidate for the first full-scale military communications satellite system. Before the final go-ahead, the Pentagon studied the idea of sharing satellites with the new Com-

munications Satellite Corporation (now Comsat). But the
Defense Department and the Corporation could never get
together on the design of the satellites or the cost. And the
corporation was going to let foreigners use the satellite: a
combination military-civilian satellite just got too complicated. In 1964 McNamara said the military could have their
own communications satellite system. Philco (later Philco-
Ford, later Ford Aerospace and Communications) got the
contract.

Meanwhile, the Air Force had been developing a new
rocket, the Titan III-C, that could carry the satellites a lot
higher than 5,000 miles; it could take them out to the
geosynchronous orbit. But the Philco satellites weren't
designed for that. They lacked the small rocket thrusters
and the electronic equipment that they'd need to 'keep
station,' to keep from drifting out of the necessary fixed
position. So Philco, Aerospace, and the Air Force came up
with a compromise: send the satellites into 21,000-mile
orbits, 1,250 miles short of the geosynchronous. There,
from an earthly point of view, they would drift eastward
about 30 degrees a day. But with enough of them spaced far
enough apart, the recurrent loss of contact wouldn't matter: as one moved out of range of two communicating
ground stations, another would move in to replace it. And
you wouldn't need the complex ground control efforts of
the Satellite Control Facility.

The IDSCP satellites were twenty-four-sided
polyhedrons, each about a yard in diameter and covered
with 8,000 solar cells for power. Stacked up together for
launches of up to eight at a time, they looked like a giant
Tinkertoy sculpture. Launching them was tricky – the upper stage of the rocket had to drop them off one at a time in
a 25,000-mile-diameter circle around the equator. But the
system worked, and between June 1966 and June 1968, the
Air Force got twenty-six into orbit. By 1967, Lyndon
Johnson could talk by military satellite to his boys in
Vietnam.

With the IDSCP satellites, eleven telephone conversations could go on at once (though only five of the quality we
are used to on our own phones) or 1,550 teletype circuits

could be set up. This translated to 1 million bits per second. The bit is the computer's 'one or zero,' its 'yes or no.' It takes about eight bits to stand for a single letter or digit. To use the system required big ground stations, for a satellite's whole solar-generated power system only produced 40 watts; its transmitter sent 3 watts of signal power. That meant that it took a powerful transmitter to send the satellite a signal it could register, and a powerful ground receiver to pick up the satellite's weak retransmission of the signal.

The system also used a radio frequency that required large antenna dishes on the ground: it received at about 8 gigahertz (GHz) and sent at about 7.3 gigahertz. 'Hertz' (from German physicist Heinrich Hertz) is now the standard term for what used to be called 'cycles per second' of electromagnetic waves. Your AM radio gets its news and music at around 1 megahertz (million hertz); FM radio receives signals at about 100 megahertz (100 million hertz); channels 2 to 13 (also referred to as VHF, or very high frequency) on your television are just above and below the FM radio band; the UHF (ultrahigh frequency) TV channels operate in a band of frequencies between 500 megahertz and 1 gigahertz (1 billion cycles a second).

The IDSCP satellites worked pretty well, but they were not geostationary. As things turned out, the first regularly operating geostationary military communications satellite didn't belong to the United States (though the U.S. did put up an experimental one in 1968 and another in 1969). And it didn't belong to the Russians, either. On November 22, 1969, the British Skynet 1 went into orbit. Actually, the U.S. Air Force (with the help of The Aerospace Corporation) wrote the specifications. Philco-Ford, builder of the IDSCP satellites, built the Skynet, using much the same electronics. NASA launched the satellite from Cape Canaveral. And the Americans also built the British control station at Oakhanger (it not only controls the Skynet, but is part of the U.S. Air Force Satellite Control Facility).

The Skynet was, of course, an improvement on the IDSCP satellites. Its antennae were 'despun' (stayed still while the rest of the spacecraft spun around it), allowing for a stronger broadcast signal. Its signals could be picked up by somewhat smaller ground antennae. It had batteries for continued power when the earth came between it and the sun. And it had small thrusters and the necessary radio links for 'station keeping' – maintaining its precise position in orbit. It worked well and several successors have followed the Skynet 1.

The next geostationary military satellite *still* wasn't officially American: it was called the NATO II (the NATO I system was actually a NATO arrangement to share some use of the IDSCP satellites already in orbit). The allies paid for it jointly, and Philco-Ford built it along the lines of the Skynet 1. The U.S. Air Force controlled the spacecraft. In the late seventies three more capable NATO III communications satellites went on station.

While all these other communications satellites were going up, the Defense Department had a spectacularly more capable spacecraft on the drawing boards: it took the name DSCS II (pronounced 'Discus Two') from Defense Satellite Communications System II. TRW got the contract. The satellite (fifteen have been made, but only ten made it to orbit; even fewer worked as intended) weighs 1,300 pounds – ten times more than the IDSCP spacecraft – but it's more than a hundred times as capable. Where the IDSCP could relay 1 million bits of digital data a second, the DSCS II can handle 100 million. Where the earlier satellite could carry up to eleven two-way voice circuits at a time, the DSCS II could carry 1,300. A cylinder 9 feet in diameter, the satellite is 13 feet high if you count its antennae. It can relay messages either across the whole disk of the earth as seen from 22,250 miles, or it can focus its signal down to a narrow circle, making it easier for smaller ground antennae to pick it up, as well as making enemy interception more difficult.

The DSCS II system (four operate full-time, one is a spare) handles military communications between fixed ground stations, using about the same superhigh frequencies (7 to 8 GHz as the original system; it is difficult for mobile military forces to carry around antennae big enough to use it. That's one reason the Navy decided it needed its own satellite system, called the Fleet Satellite Communications System (FLTSATCOM, pronounced 'Fleetsatcom,' for short). TRW built those satellites, too. FLTSATCOM satellites send in the UHF part of the spectrum, in the area of 200 to 400 megahertz (somewhere between your FM radio and your UHF television reception). One FLT-SATCOM channel is assigned to the Fleet Broadcast Communications System. Big 60-foot-diameter ground antennae send messages up at superhigh frequency (at about 8 GHz), and the satellites broadcast the messages around the world. All U.S. Navy ships at least carry receivers for these messages, and there are satellites that span Southeast Asia to the West Coast of the U.S., Hawaii to the East Coast, the East Coast to the Mediterreanean, and Africa to the Philippines.

Many ships carry two-way radio equipment in order to use the satellites, allowing them to talk both with Navy shore bases and with other ships. Each satellite has nine relay channels for that purpose. Each also has a very-wide-band channel (500 kilohertz) for general Defense Department use. Twelve more channels go to the Air Force.

As capable as DSCS II and FLTSATCOM are, they're not enough. Before the first FLTSATCOM satellite went up in February 1978, the Navy used channels on Comsat General Corporation's MARISAT; this service, dubbed GAPFILLER to underscore its temporary nature, has continued even though all five FLTSATCOM satellites are in orbit. There's one MARISAT satellite each over the Atlantic, Pacific, and Indian oceans. The Defense Department rents six channels on each satellite, mostly for the Navy Fleet Broadcast System and two-way communication with ships, but partly for Army use as well.

In 1977 Congress told the Navy it had to stop buying the FLTSATCOM satellites, that for its next generation of communications satellites it was to rely on leasing services from private industry. The admirals didn't like it, but they had no choice. In 1978, after receiving bids, the Navy awarded a contract to Hughes Communications Services, Inc , for LEASAT. For $353 million, Hughes would provide the services of four geostationary satellites for five years. The satellites would ride into space on the space shuttle and the Boeing IUS. The satellite system would have thirteen channels where FLTSATCOM had twenty-three.

In mid-1981 the Navy went back to Congress to ask for both revisions in the Hughes contract and permission (and money) to start building more FLTSATCOM satellites. Hughes in 1978 had based its planning on the premise that the shuttle would be available in time to get the first LEASAT operating by April 1982. The contract said Hughes would start getting its money when it started giving service, but with the delays, Hughes was in money trouble – its 'financial exposure' was a lot bigger than foreseen in 1978. Early in 1981 the Navy went to Congress for a $59 million advance for Hughes in 1982.

One additional FLTSATCOM satellite in orbit would cost the Navy another $200 million (less per satellite if more are bought), but the Navy thinks it's worth it. As the military demand for UHF satellite communications channels got up in the mid-1980s, the available capacity would go down with sole reliance on LEASAT. In June 1981, Vice Admiral Gordon R. Nagler, the Navy's command-and-control director, told the House Armed Services Committee: '. . . I want to state at the outset that all Department of Defense satellite communications requirements, including those required for the Rapid Deployment Force, will not be satisfied through 1989 with our existing and scheduled satellites, nor even with your authorization . . . [of the Navy's requests]."

Today the U.S. military establishment is a million bodies smaller than it was in 1970. The U.S. has fewer foreign bases and fewer allies than it did then. Why this burst of demand for military communications channels? Part of the demand has to do with the loss of earlier bases – which had served as communications relay stations – in such places as Ethiopia and Morocco. Admiral Nagler, in his written statement to the Congressmen, said: 'A return to non-satellite communications would require remanning facilities as international political situations would allow and would create heavy backlogs in message and data processing because of manual handling delays [e.g., the *Pueblo* incident]. The reconstitution of facilities and manpower to support operations such as in Southwest Asia could be so time-consuming as to effectively nullify many options available to the United States.'

Then there's the increased reliance on high-speed, high-volume data transmission for handling the military's enormous personnel management, pay, and logistics costs. Nagler said the work loads in these areas would be 'extreme' without satellites. He also said that the Pentagon would be less able to carry out plans to 'expand the use of computer-to-computer weapons command and control and futher reduce manpower requirements while increasing the speed and efficiency of response and management . . .' The expansion of the electronic battlefield will need satellites.

These sources of increased military demand for satellite communications are important, but perhaps not the most important. Two new tasks the Pentagon has taken on in the past few months explain a considerable part of the demand. One is to be able to intervene almost anywhere around the world, at a moment's notice, with troops based in the United States. The other is to be able to fight a *long* nuclear war. Both tasks may be impossible, but the Pentagon is going to try.

In 1979 the Carter Administration approved a Pentagon plan for an Army Ground Mobile Forces satellite communications program. At about the same time the Administration decided to put together a Rapid Deployment Force (RDF) for the quick movement of U.S. military forces to faraway places. As Secretary of Defense Harold Brown explained just before he left office in 1981, 'While the potential missions of our Rapid Deployment Forces are global, in practice most of our planning and programming has focused on Southwest Asia.'

The mission? Protect Persian Gulf oil.

Brown pointed out that 'aside from political complexities within the area, distance is the central problem we confront in landing Rapid Deployment Forces that can defend our interests in the vital Persian Gulf–Indian Ocean region. By air from the East Coast, Southwest Asia is over 7,000 miles away; by sea through the Suez Canal, about 8,000 miles; and by sea if the Canal were closed, over 12,000. To a large extent, distance drives our plans and programs . . .' Many kinds of preparation are going into this capability for long-distance intervention: building new airplanes, finding places in the area to store supplies in advance, practising joint operations like those the U.S. has held with the Egyptians. But now that the U.S. no longer has communications stations in either Iran or Ethiopia, all the rest would be much more difficult, perhaps unthinkable, without satellite communications.

The Pentagon already has communications to the Indian Ocean by means of satellite terminals on the island of Diego Garcia and on the Navy ships in the region. But the plan is to link the mobile ground forces, Army and Marines, into the satellite communications net as well. As the civilian chief of Communications in the Pentagon, Gerald Dineen, said in 1980, 'The satellite communications technology and the equipment we now have in the field have completely changed the way we are able to do our military defense business.' In 1979 the Army (which is the executive agent for all the services for all ground satellite terminals) started a multiyear program of buying 226 new ground-mobile satellite terminals for Army use.

Of these terminals, a major type, the AN/TSC/85A, will let the Army hook up with the SHF (superhigh frequency) channels of the DSCS II satellites. The terminal will ride on two small trucks. One truck will carry a dish antenna about 8 feet in diameter; the other truck will carry the electronic equipment and operator's station. The trucks are small enough to ride on the C-130 and C-141 aircraft of the Military Airlift Command. By using SHF, that terminal will be able to set up several circuits at once between division command posts and their maneuvering brigades (as well as linking them all with distant headquarters or even the Pentagon), and it will be able to use special techniques to fend off enemy jamming attempts.

A second new type of terminal will fit on one truck and

will plug into the UHF channels of the FLTSATCOM satellite and the LEASAT. With it, lower-echelon forces will be able to reach each other and coordinate their movements, even at long distances, just as Navy ships now can.

A third type of terminal is the AN/PSC-1 manpack. It's designed for Army Special Forces and Ranger units, and one man can carry it. The electronic equipment will transmit messages in high-speed bursts of UHF signals, keeping the enemy from detecting and tracking down the location of the sender. Small units would take the manpack terminal on secret missions behind enemy lines. (The Air Force is working on a terminal small enough for an attaché case. The operator holds the antenna in one hand and a telephone handpiece in the other.)

Satellite communications don't necessarily make Tolstoy's conclusions obsolete. The flaw in the picture of Napoleon with satellites is that although satellite communications technology has advanced 'command and control' dramatically, other military technology has made the military commander's problems proportionately greater. Communications move faster, but so do battles. Where Napoleon's fastest soldiers traveled as fast as a horse can run, today's U.S. Air Calvalry moves by helicopter. Jet planes and rockets exceed the speed of sound. A great-power war today is fought not on a single battlefield at a time, but on many battlefields across thousands of miles. Putting aside, for the moment, the question of nuclear weapons, the firepower of today's weapons is hundreds of times greater than the muskets and cannon of Napoleon's time.

The satellite-assisted military commander might be hampered as greatly by too much information as Napoleon was by too little. Computers can help, but how could any one man keep the whole picture of a modern battle in his head? The vast scale, the speed, the intensity, the complexity of modern warfare, may replace mere bad information as the bane of true 'command and control.' Yet, the military planners don't think so, and their answer is to build ever-more-sophisticated 'C-cubed' technology.

But what about *nuclear* war? Surely and exchange of nuclear weapons would make the Battle of Borodino look like a Sunday-afternoon football game. Utter physical devastation. Millions dead and dying. Economic and social fabrics shredded. Battlefields and cities alike wastelands of lingering radioactivity. What does command and control mean under those circumstances? The military planners aren't giving up here, either: one of the latest enthusiasms at the Pentagon is to improve command and control of the nuclear forces. 'Strategic connectivity,' they call it.

When Lewis Carroll's Alice stepped through the looking glass, she entered a world with a logic – or illogic – all its own. Right now, as you read this, an airplane called 'Looking Glass' circles over the central United States. That Looking Glass, too, is a portal into a world with a logic all its own, a logic foreign to our daily experience: the logic of nuclear deterrence. Looking Glass is the code name for the Air Force Strategic Air Command's airborne command post. There is actually a fleet of such planes, but one is always in the air: it doesn't land until another takes off. The logic of the Looking Glass world is expressed in the SIOP, the Single Integrated Operational Plan – the U.S. plan for nuclear war. The mission of Looking Glass is to pass on weapons to our ICBMs and bombers to deliver their nuclear weapons onto the Soviet Union.

Looking Glass is just one link in the nuclear command-and-control chain, one 'segment of the strategic C-cubed architecture,' as the military bureaucrats might put it.

Imagine that tomorrow, without notice, the Soviets decide to relieve themselves of the American opposition to their international policies by destroying us as a military power. The U.S. early warning satellites over the Eastern Hemisphere detect the launch of Soviet missiles and their information is transmitted (by satellite) to the North American Aerospace Defense Command inside Cheyenne Mountain, Colorado. The Western Hemisphere warning satellites spot submarine-launched missiles at about the same time. Moments later, the radar curtains around the United States and extending from Clear, Alaska, to Fylingdales, England, confirm that missiles are on the way.

The Command Center at NORAD has already sprung into action. One of its Honeywell 6000 series computers is at work calculating how many missiles are headed where. Another manages the flow of information in and out of the mountain, sending the NORAD calculations to three other places: to the Strategic Air Command headquarters at Offutt Air Force Base, Omaha; to the National Military Command Center underneath the Pentagon; and to the Alternative National Military Command Center near Fort Ritchie, Maryland (about five miles from Camp David). Those three centers also receive direct readouts from the satellites and radars that watch for submarine missile launches (but not from those that look for ICBMs or bombers).

The duty officers at NORAD and the other three command posts call a 'Missile Display Conference' to discuss the information coming in. Within a couple of minutes, they decide that this could be the real thing and call their superiors, moving to the stage known as the 'Threat Assessment Conference.' At this stage, the Strategic Air Command has ordered its alert bombers to get ready to take off, before submarine-launched missiles, a few minutes away, can reach them. In Hawaii, the Commander in Chief, Pacific's airborne command post has taken off. The airborne command posts of the other 'nuclear cincs,' Commander in Chief, Atlantic and Commander in Chief, Europe, are also soon aloft.

Within five minutes of the moment they were warned, the B-52 and FB-111 bombers have taken off (before they approach Soviet airspace, they will turn around and come back unless ordered otherwise). Minutes later, the four ground command posts have notified the White House Communications Center of their assessments. The President joins in a 'Missile Attack Conference' to decide what to do next. Persuaded that an attack is under way, the President boards his helicopter for Andrews Air Force Base, just outside Washington. He, the Secretary of Defense, and the Joint Chiefs there climb aboard a waiting Boeing 747, known formally as the Nationally Emergency Airborne Command Post (NEACP), known informally as 'Kneecap.'

The reason for all these airborne command posts is the working assumption that all the ground command posts – the Pentagon, the bunker at Fort Ritchie, the Strategic Air Command underground posts at Offutt and Cheyenne Mountain – will be demolished about thirty minutes into the war. In this age of missile accuracy, the only safety is in mobility.

Even before his plane is in the air, the President has to make a hard decision: does he order the Strategic Air Command to launch its Minuteman Intercontinental Ballistic Missiles out from under the Soviet attack or does he wait to find out where and when the Soviet nuclear weapons will explode? In theory, a high percentage (some say around 90 percent by the mid-1980s) of those missiles are at risk of being destroyed in their underground silos, which can no longer protect them from the highly accurate Soviet warheads. At the same time, others say that the chances of perfect Soviet machinery executing a perfect attack to get the theoretically projected results are slim indeed.

Suppose the President decides to launch. The military always with him pulls out the 'go-codes' – the secret messages that will verify to the military commanders that this is really the President and that he is really authorizing release of the Emergency Action Message. From the NEACP, possibly via satellite, the message goes out. Now Looking Glass goes into action: under missile fields in Missouri, North Dakota, Montana, Wyoming, Arizona, Arkansas, missile launch control centers receive their orders. They're all equipped with satellite terminals in case their other communications links are cut.

Overhead cruise other SAC (Strategic Air Command) aircraft, parts of the Airborne Launch Control System. If the underground launch control centers are destroyed, these aircraft can take over the launching of the missiles below. If communications to any of the bases have broken down, the NEACP may order the launch of certain Minuteman missiles which contain the tape recorders and radios of the Emergency Rocket Communications System, ERCS. The commanders can record the Emergency Action Message on the ERCS, and the rocketborne radios will play the message back from hundreds of miles up.

While SAC is launching the missiles and passing final target orders to the bombers, the Navy orders its submarines to prepare to launch their nuclear missiles. The Navy has ground-based low-frequency and very-low-frequency radio broadcast stations, but these too have to be considered expendable. Always in the air over the Atlantic is a TACAMO plane (whatever its originals, TACAMO has come to signify, 'Take Charge and Move Out' in Pentagon parlance). This plane trails a very long wire antenna to broadcast the Emergency Action Message via very-low-frequency radio to the submerged ballistic missile submarines. (Both the TACAMO and the subs carry satellite UHF communications equipment as well.) Many of the subs (about seventeen will probably be at sea, with about 180 missiles and 2,000 to 2,500 nuclear warheads among them) will be instructed to stand by for later launch orders – perhaps days or weeks later. Others will launch missiles immediately, so as to 'soften up' Soviet air defenses for the approaching B-52s, FB-111s, and air-launched cruise missiles.

It would seem to be all over at this point. Nuclear war. National destruction. The survivors envy the dead. As the military might put it, 'Deterrence has failed.' Our bombers and subs may wreak their revenge, their 'retaliatory strikes,' but they'll have little to come home to.

But such is not the thinking in current military doctrine. According to General Richard Ellis, USAF (Director, Joint Strategic Connectivity Staff, Joint Chiefs of Staff; Commander in Chief, Strategic Air Command; Director, Strategic Target Planning, Joint Chiefs of Staff): '. . . we have come a long way since the early days of nuclear planning when we were concerned only with the initial attack. We designed our command and control system to launch the force in a massive strike. Today, for all practical purposes, that is the connectivity that still exists, with some refinements.

'In those days, we were only worried about getting the message out. We were not worried about survival of that system. Today, our nuclear strategy has changed under national directive to the point where we are required to have a flexible plan, to have options available to the President, to have an enduring capability that can last for an indefinite period, and to be able to exercise control over a reconstituted force after perhaps several exchanges.'

Ellis saw 'strategic connectivity' as a continuous loop of information and action. It doesn't end with the execution of the orders in the Emergency Action Message: 'Once execution is complete, we must have sufficient feedback data and intelligence information to evaluate not only the effects of our retaliatory attack on the enemy but also our own resources remaining to continue the war. . . . Finally, we need communications in the trans- and post-attack period to reconstitute our surviving forces, generate follow-on sorties and retarget our forces, as required. This complicated cycle is then repeated until hostilities are terminated.'

That's a big order. 'It is clear,' said Ellis, 'that a communications network, survivable throughout the entire spectrum of conflict, is necessary.' As a first step toward building such a network, the Air Force has put together AFSATCOM, the Air Force Satellite Communications Network. AFSATCOM has no satellites devoted solely to its purposes – its transponders (relay radios) ride on other satellites. One type of satellite with AFSATCOM equipment is the FLTSATCOM. It has twelve UHF channels dedicated to the system.

Above about 70 degrees north latitude, it's hard to make contact with satellites that circle the equator. So the Air Force has other satellites, the Satellite Data System (SDS), for polar communications. The SDS satellites travel in what is sometimes called a Molniya orbit, after the series of Soviet communications satellites that first used it and still do. The Molniya orbit is highly elliptical: in its nearest pass to the earth, its perigee, its altitude is only about 200 miles; but at the farthest end of its swing, its apogee, it is more than 24,000 miles away. That gives the orbit a period of twelve hours, and most of the twelve hours is spent in the higher-altitude part of the orbit (the farther from the earth, the slower a satellite travels). The plane of the SDS orbit is inclined about 62.5 degrees to the equator. Since it goes around the world twice a day, half the time its high-altitude loop is over North America and the other half it is over the Soviet Union. (The Air Force Satellite Test Center – the Big Blue Cube in Sunnyvale – also uses the SDS to contact its Remote Tracking Station in Thule, Greenland; in addition, the satellite may play a special role in the control of spy satellites.)

Besides having AFSATCOM terminals at ground bases and on the airborne command posts, the Air Force is in the process of putting them on every strategic bomber and tanker, on its major electronic warfare planes and reconnaissance planes, and at every missile launch control center. (By 1986 the Air Force will have spent $1.5 billion on 920 such terminals.) Using AFSATCOM, a bomber en route to the Soviet Union should be able to get last-minute target instructions from Looking Glass or from Kneecap. A bomber on its way back should be able to report whether it reached its target. A reconnaissance plane should be able to give on-the-spot assessments of the damage U.S. missiles may have done to Soviet targets.

Besides its American-based bombers, submarines, and missiles, the U.S. has more than 200 nuclear weapons storage sites in Europe (and a few other places in the world). Those sites too are plugging into the satellite communications network: all have or will soon have MSC-64 UHF satellite terminals. John Morgenstern, a former Pentagon director of Theater and Strategic Command and Control, explains the trend: 'The evolution of doctrine for the use of tactical nuclear weapons in Europe has a parallel in the evolution of the doctrine for the use of our strategic forces. In the strategic area, we have moved from the idea of a single, massive exchange through the era of flexible response to the idea of fighting a nuclear war over a period of months. In the tactical arena in Europe, we have moved from the trip-wire philosophy . . . toward the idea of fighting a war with conventional forces and employing tactical nuclear forces (TNF) when needed in a flexible manner as an integral part of the fighting. . . . What is evolving is a doctrine of war fighting in the conventional sense but including the use of tactical nuclear weapons.' This doctrine, he says, requires 'survivable' command and control mechanisms – like satellite communications.

Just how plausible is the idea of a 'limited' or 'protracted' nuclear war? In his final report to Congress, Secretary of Defense Harold Brown said, as he had many times before, '. . . I remain highly skeptical that escalation of a limited nuclear exchange can be controlled, or that it can be stopped short of an all-out, massive exchange.' Even so, he went on, '. . . I am convinced that we must do everything we can to make such escalation control possible, that opting out of this effort and consciously resigning ourselves to the inevitability of such escalation is a serious abdication of the awesome responsibilities of nuclear weapons, and the unbelievable damage their uncontrolled use would create, thrust upon us.'

In Brown's version of 'escalation control,' we would try to persuade the Soviets in the early stages of a nuclear war that going on would bring them more losses than gains. Therefore, we should at the beginning leave them with valuable targets still unhit, but at risk. Brown's idea is to end the war as quickly as possible – if possible. And, he added, 'The key to escalation control is the survivability and endurance of our nuclear forces and the supporting communications, command and control, and intelligence [$C^3I$] capabilities.' If the Soviets know we have these things, then they are more likely to be deterred from attacking in the first place.

Brown emphasized that the new preparations for 'nuclear war fighting' that he was promoting did not mean he thought we could 'win' a nuclear war, but only that he wanted to persuade the Soviets that *they* could not win one. It's not clear that all the uniformed military men are quite so pessimistic. For example, Lieutenant General James W. Stansberry, head of the Air Force Electronic Systems Division, was recently quoted as saying, 'In previous years the concept for C-cubed was that it only had to be able to get off a launch of U.S. strategic weapons in response to a first strike before damage was unacceptable. The idea that there was no way to win a nuclear war exchange sort of invalidated the need for anything survivable. There is a shift now in nuclear weapons planning, and a proper element in nuclear deterrence is that we be able to keep on fighting.'

By the logic of the Looking Glass world, the logic of the SIOP, the best way to avoid nuclear war is to be convincingly ready to fight it. A Lee Lorenz cartoon in *The New Yorker* has one general say to another, 'As I see it, our commitment to the peace process is only credible if our commitment to the war process is credible.' That expresses the logic of deterrence exactly. It is not enough that the Soviet Union should be threatened with unprecedented damage to its society and economy in a nuclear war: Soviet military planners, the logic goes, are more likely to be deterred from risking war if they believe that after the first round, they will have fewer nuclear weapons left in reserve than the U.S. does. They should be made to believe that no matter what else happens to our society, we will be able to continue lobbing nuclear weapons back and forth as long as they can; that no matter how much damage they can inflict on the United States, they will suffer even more damage themselves; that no matter how many decades it might take the United States to 'recover' from a nuclear war, it would take them even longer.

This modern logic of deterrence denies that there is some absolute level of destruction that the Soviets might suffer in a nuclear war which would outweigh any conceivable gains they might hope to achieve in starting one. Rather, it says that the calculus of deterrent is relative – that projected game counters like numbers of 'residual' weapons, percentages of industry left, percentages of population destroyed will determine winners and losers. And if the Soviets add up the projected counters and it looks as if they come out ahead by those measures, they might be more willing to go to war. But trying to make the projected score look right is a much harder task than just trying to guarantee terrible punishment. It calls for a sophisticated nuclear war – fighting strategy. 'Strategic connectivity' would be vital to such a strategy.

Building satellite communications systems that can survive a nuclear war is going to be a demanding task. The satellites themselves might come under Soviet attack. The Soviets don't have an anti-satellite weapon today that can reach out to the geosynchronous orbit, but U.S. military planners want to anticipate the possibility. In the meantime, satellites are vulnerable to the effects of high-altitude nuclear explosions even at very great distances. In particular, nuclear explosions emit intense bursts of energy across the electromagnetic spectrum – the so-called EMP, or electromagnetic pulse effect. The EMP effect can damage, temporarily or permanently, all kinds of electronic equipment, in space or on the ground, and could play havoc with satellite communications systems. And, of course, the ground segments – terminals, switching stations, land lines – of military satellite systems will be subject to the destructive blast and heat effects of nuclear weapons.

So Pentagon planners, particularly those in the Air Force Space and Electronic Systems divisions, are busily trying to figure out ways to cope with the threats to 'survivable and enduring C-cubed.' Some of the measures they have come up with are already in place, some have been the subject of experimentation, some are still on the drawing boards.

The FLTSATCOM satellites, for example, are 'nuclear hardened' against the electromagnetic pulse effect, as are the AFSATCOM terminals that plug into it. One reason the Navy and the Air Force don't like the LEASAT as a sole replacement for the FLTSATCOM-AFSATCOM satellites is that it isn't EMP-hardened. According to General Randerson of the Strategic Air Command, '. . . this [FLTSATCOM] is the only EMP-hardened system we have. The space segment is hardened, the airborne segments are hardened, and we are going to harden against EMP the launch control centers for our missile force and for the nuclear weapons' storage sites. This is the only satellite system that is really a war machine and so FLTSATCOM is very dear to SAC's operations.'

The next generation of Defense Satellite Communications System Satellites, DSCS III, is also EMP-resistant.

General Electric has designed the satellite to resist not only the effects of nuclear weapons, but also the effects of Soviet radio jamming attempts. Its superhigh-frequency (SHF) receiving antennae can focus down on the source of a message, automatically 'nulling' interfering radio noise from other directions. Its transmitting antennae can focus the signals (from six separate SHF channels) down into narrow beams covering circles on the earth just one degree (about 69 miles) across, making enemy interception or jamming much more difficult.

The DSCS III will also carry a special Single Channel Transponder for relaying the Emergency Action Message. It will receive the nuclear-go messages on either SHF or UHF, then rebroadcast them to all the AFSATCOM terminals. The satellite is designed to do much of its own station-keeping (maintaining its orbit) and attitude control (keeping its solar cells and antennae properly aimed), so that if the Satellite Control Facility ground stations were destroyed, the satellites could keep working. By the mid-1980s four DSCS III satellites should be posted over the Atlantic, East and West Pacific, and Indian oceans.

As sophisticated as the DSCS III spacecraft is, it doesn't satisfy military demands for 'survivable, enduring' communications. Other ideas are in the works. The MIT Lincoln Laboratory incorporated some of these ideas in a pair of satellites that the Air Force launched in 1976. The Lincoln Experimental Satellites, LES-8 and LES-9, can relay radio signals not only from one point on the ground to another, but from one point on the ground to one satellite to another satellite to another point on the ground. With this 'crosslink' between satellites, only two satellites can provide communications service to three quarters of the earth's surface. The main frequency band is the UHF used by the other 'tactical' communications satellites, but the satellites can also send and receive in the EHF (extremely high frequency) range, at 38 gigahertz – unprecedented in satellite communications. Such frequencies would transfer data at much higher rates than those now in use, and they would be extremely difficult for an enemy to jam.

The solar cells that most satellites use for power are

subject to various kinds of radiation damage, say from atomic weapons or lasers. Instead of solar cells, LES-8 and -9 use small nuclear power plants of a sort – 'radioisotope thermoelectric generators' that make electricity from the heat of decaying radioactive elements. The satellites also carry an experiment, designed by the Charles Stark Draper Laboratory in Cambridge, to test a gyroscope for a self-contained attitude control system. The idea, again, is to make satellites independent of vulnerable ground stations.

For three years running, the Pentagon went to Congress with a proposed satellite that would apply several 'survivability' ideas in an 'operational,' not just experimental, system. From 1978 to 1980 the generals tried to sell Congress on 'STRATSAT,' or the Striegic Satellite System. In 1980 Secretary of the Air Force Hans Mark explained why he thought the STRATSAT was worth spending more than $2 billion on between 1981 and 1997: 'I think the judgment . . . has to be made on the basis of how important you believe survivability after a nuclear exchange is. It is my understanding that the Soviets do, in fact, think about what happens after a nuclear exchange, that they are willing to contemplate that kind of thing happening. I would regard the investment in such a satellite system as an investment in something that surely would survive a nuclear exchange.'

The proposed satellite wouldn't have been in an equatorial orbit, but one that went over the poles, at *five times* the altitude of geosynchronous satellites – about 111,000 miles, almost halfway to the moon. They would be out of range of the effects of nuclear weapons exploded near the earth, and certainly out of range of Soviet anti-satellite weapons, which can't even reach the geosynchronous altitude. Operating in a constellation of four, the STRATSATs would be crosslinked by EHF (extremely high frequency) radio in the band between 55 and 65 gigahertz, a frequency not yet used in satellite communications. The military are interested in moving more communications into the EHF bands both because they could move greater masses of data more quickly and because they could apply their anti-jamming techniques more effectively. For links with planes, ships, or ground units, the STRATSAT would have UHF (like the FLTSATCOM, SHF (like the DSCS satellites), and EHF channels.

In the end, the Pentagon failed to persuade the Congressmen to spend $3.5 billion on a satellite whose sole purpose was to maintain communications during nuclear war. After the third year of rejection, the communications planners dropped STRATSAT, went back to their drawing boards, and came up with Milstar. The acronym comes from Military Strategic–Tactical and Relay, and the idea is to combine more communications services in one type of satellite. Unlike STRATSAT, Milstar will probably be in geostationary orbit; like STRATSAT, it will be designed to survive a nuclear war and provide 'strategic connectivity' to the nuclear forces. Major General Gerald Hendricks, vice commander of the Air Force Space Division, boasts, 'Milstar is designed to be a war-fighting system. The first of its kind.' He says it will work '. . . during all levels of conflict, have worldwide two-way communications, and be survivable and enduring.' The system would begin the military move to extremely high frequency, perhaps around 44 gigahertz, but it would also keep UHF channels because so many terminals around that use the lower frequencies. To satisfy Congressional criticisms, Milstar would serve other military users besides the nuclear forces. It would probably be the successor to the Navy FLTSATCOM-LEASAT combination.

The trend in military communications is toward using higher and higher radio frequencies, with EHF – tens of billions of cycles per second – the next step. But the trend doesn't stop there: the Navy and the Air Force are now working on light, in the form of lasers, as a means of communication. In the electromagnetic spectrum, EHF frequencies are in the tens of billions of hertz; light-wave frequencies are in the hundreds of thousands of billions. The Air Force Space Division and McDonnell Douglas are working on a satellite laser communications system that could transmit the entire contents of the *Encyclopædia Britannica* in just over a second, or relay 250,000 separate telephone conversations at one time.

With a laser communications system in place, the Air Force could send a message from an airborne command post like Looking Glass, flying over the United States, up to one satellite in a geostationary orbit, across to another satellite, and down to a bomber on the other side of the world. Vulnerable ground relay stations would be unnecessary. And the transmitting beam from the laser would be so narrow – about half a mile across at the earth's surface – that it would be almost impossible for the enemy to intercept or jam it.

The Navy is interested in laser communications for its submarines. Radio waves don't penetrate water well. To get a UHF message from a satellite, a submarine has to have an antenna on the surface. To get a very-low-frequency message from a Navy shore station or from one of the TACAMO airplanes, a submarine has to trail a buoy near the surface. Having to approach the surface may make it easier for Soviet anti-submarine forces to find the Navy's ballistic missile submarines before or during a war. But blue-green laser light *can* penetrate deep underwater. So the Navy has given contracts to TRW, GTE Sylvania, Lockheed, and McDonnell Douglas to work out ways to communicate with the subs by laser.

The Navy Electronic Systems Command has been interested in two basic schemes for getting the messages out. In one scheme, the Navy would put a big mirror in geosynchronous orbit. A land-based laser would aim its signal at the mirror, which would reflect it back down to the submarine. In the other scheme, the laser itself would be aboard the satellite. Ground-based radio stations would send the messages up, then the satellite would transform them into laser signals and beam them down to the sub. In either case, the ground stations could be made mobile, to increase their chances of survival once a nuclear war had begun.

Pentagon plans for maintaining 'strategic connectivity' during nuclear war also have to take into account that Soviet anti-satellite weapons could destroy U.S. communications satellites – a feat that's not possible now, but that might be in years to come. One way of dealing with that problem would be to try to put up some kind of replacement satellite after the big geosynchronous ones are destroyed. The likely launch pads would be submarines, since land-based launching sites would have been eliminated by Soviet bombs. One kind of replacement would be a low-altitude (say 600 to 1,000 mile) 'store-dump' satellite. As it passes over the sender, it picks up and records the message; later, when it passes over the receiver, it plays the message back. (As it happens, the Soviet Union has been operating a system like this for some years to communicate with its Navy: it launches the satellites in packages of eight in a single launch, and probably has three sets of eight operating at one time.)

Another idea the Defense Advanced Research Projects Agency (DARPA) is studying is called PACSAT – Passive Communications Satellite. The Pacsat would be a long string of shiny beads. In a low orbit would be a 3,000-foot chain of thousands of small spherical reflectors, less than an inch across. Although Soviet radars or telescopes wouldn't be able to spot the beads, U.S. transmitters on the ground or in the air could bounce radio messages off them, sending out Emergency Action Messages to the nuclear forces.

Ideas like PACSAT are the responsibility of the Strategic Technology Office at DARPA. Colonel Charles Heimach, USAF, was Assistant Director for Advanced Concepts in that office until the fall of 1981, when he began a year of study at the National War College. Heimach has been working on space programs since he entered the Air Force in 1960. He was for a time the chief launch controller for the U.S. anti-satellite missile based on Johnston Island in the Pacific. He has been working on anti-satellite programs, the surveillance of space, and the defense of satellites ever since.

For the spacemen's convention at the Air Force Academy in 1981, Colonel Heimach prepared a paper called 'Space Survivability – A Philosophy/Police Argument.' There he argued that, with appropriate preparations, U.S. communications and other satellites at high altitudes could survive nuclear war – the 'execution of the

SIOP.' In fact, he wrote, it would be *easier* to protect those satellites in the 'post-SIOP' world because, for one thing, we could have destroyed the Soviets' land-based anti-satellite launchers.

In an interview he said, 'For many years people felt – and this may be true – that the execution of the SIOP would be that everything was thrown at each side and the world came to an end, and so you really didn't care about the satellites. Well, there's a change in thinking nowadays that says that's not necessarily so. If you read Soviet doctrine, they clearly plan in *all* their wars to conduct a protracted type of war until they bring the other side completely under their domination. So people are starting to rethink what might happen in a war. The problem is, you can't sign up to one scenario: the worst can happen in many different ways, and you have to prepare to deal with it.'

The nuclear strategists are attempting to give the President – or the surviving National Command Authority – a variety of nuclear options from which to select. The scenarios for nuclear war include: limited 'demonstration' attacks to 'show resolve' and get the other side to back down in a crisis; 'counterforce' attacks aimed mainly at missile silos, bomber fields, and missile-launching submarines; attempts to cripple enemy responses by wrecking his command and control networks; attacks aimed at political leadership; attacks on limited sectors of the economy, such as the petroleum industry; repeated, long-term attacks designed to delay economic and social recovery.

Colonel Heimach admits that the problem with all this discussion of strategic options is that with almost any attack by the Soviets – perhaps ten nuclear warheads, perhaps a hundred – '. . . the society you and I know just doesn't exist . . . communication drops, the electrical grid becomes impossible and goes off the line . . . the population will panic.'

Since nuclear war is so horrible, he says, we need two strategies. 'You should at least have one that tries to prevent the Soviets from *ever* thinking they should execute that SIOP; but, on the other hand, you should be prepared for the full-out exchange and the protracted war, the one that lasts six months, that lasts two years or three years, and the other guy should know that you're prepared to fight these types of wars. Then when he decides to execute his SIOP . . . he'll think twice – even though we all know that any execution of the SIOP is probably going to change the world to such a degree that it's not . . . worthwhile.

'So that's why you end up speaking out of both sides of your mouth at one time – or I do: we're just trying to convince the other guy that he should never do it because we can bring the war to him.'

# CHAPTER FOUR

# Open Skies: Spy Satellites

In the rough-and-tumble days of the First Cold War, before the rise and fall of détente, the Superpowers invented the Disarmament Game (a phrase coined by Swedish author Alva Myrdal). The prize is public opinion, domestic and foreign. The game goard is usually the diplomatic negotiating table, often at Geneva. The game move is to lay a disarmament agreement proposal on the table. The object is to design your proposal to make your opponent look bellicose and recalcitrant, yourself pacific and accommodating. The favored stratagem in the game is the 'Joker.'

The Joker is the item in the player's disarmament or arms control proposal that the other side is sure to reject. The player's proposal can be unacceptable even to members of his own team (say, the Joint Chiefs of Staff), as long as it has a Joker. The Joker guarantees that, first, the disarmament or arms control plan proposed will never have to be carried out, and, second, that the onus for its rejection will go to the opponent.

In the 1950s the United States and the Soviet Union played a couple of surefire Jokers. One was the 'conventional-nuclear' Joker: the Soviet Union always proposed that nuclear weapons be abandoned first, to be followed at a later stage by troops and tanks and guns; this was always unacceptable to the United States, which saw such proposals as leaving the numerically inferior NATO forces at the mercy of the Warsaw Pact hordes. The United States, for its part, usually proposed the disarmament of conventional forces first, to be followed only in the final stages by nuclear disarmament; the Soviet Union understandably saw this as an attempt to disarm the Soviet Union while leaving the vastly superior American nuclear threat intact until the end.

The other foolproof Joker in the Disarmament Game was the 'inspection first (or last)' ploy. The Soviet Union would propose that disarmament proceed through one or more stages first, to be followed later by 'strict international control' verifying that the promised steps had been carried out. The United States was certainly not going to take the Soviet's word for it that they were disarming. United States proposals, then, usually suggested that 'on-site' inspection of military forces take place before and during disarmament; only then would everyone have confidence in the process. To the Soviets, inspection before disarmament meant opening up Soviet military secrets to Western spies.

The Soviets could always – or almost always – be counted on to reject proposals like that. Once, though, the Soviets didn't play. In April 1955 the British and French (with American approval) laid on the table a disarmament plan that moved nuclear disarmament up into the final stages of conventional disarmament, while providing for the early establishment of inspection and control provisions. A month later, to everyone's surprise, the Soviets accepted the new plan almost entirely. The shocked U.S. delegation immediately called for a recess of the disarmament talks until after a planned U.S.-Soviet summit meeting that summer. (The following September, the United States announced that it was "placing reservations" on all its presummit negotiating positions: the new plan was dead.)

Meanwhile, between the Soviet acceptance of the Western plan and the July summit talks, the United States needed a new proposal to regain lost propaganda points. A special White House committee of 'cold-war strategists,' headed by Nelson Rockefeller, came up with a plan for inspection *without disarmament*. At the summit conference President Eisenhower laid out the U.S. 'Open Skies' proposal: to reduce the chances of surprise attack, the Soviets and the United States would exchange detailed information about their military forces and then verify the information with reconnaissance aircraft and inspectors on the ground. This, said the Soviets, was "nothing more than a bald espionage plot." At the time, the Soviets had at most a handful of bombers that could reach the United States, while the United States had the Soviets literally surrounded with bombers. 'Open Skies' would tell those bombers just where the best targets were.

The fascinating thing about the Open Skies was that although the Soviets would have none of it, Eisenhower carried it out anyway: in 1956, CIA U-2 spy planes started flying deep into the Soviet Union. From 13 miles up, aerial cameras captured images of Soviet rocket test centers, bomber bases, nuclear test sites, submarine pens. The Soviets remained powerless to stop these incursions for four years. Then in May 1960, they finally shot one down. The Open Skies over the Soviet Union closed up – for all of three months. On August 11, 1960, the Air Force recovered the first film capsule from the world's first photographic reconnaissance satellite, Discoverer 13. Today, visitors to the National Air and Space Museum satellite hall amble past the Discoverer 13 re-entry pod without quite realizing what a piece of history it is.

In the early years of Soviet-American space competition (and perhaps still today), the emphasis was on the big, the spectacular, the powerful. How big are Soviet rockets? How many missiles do they have? How many million pounds of thrust can we build into a rocket (the Saturn V, which sent men to the moon, had five gargantuan engines with 1.5 million pounds of thrust each). How heavy a satellite can they put up? How long can we keep a man in orbit? Who can get to the moon and back first? But many of the achievements in space technology have not been in the realm of the mighty and the spectacular, but instead in the domain of the tiny, the intricate, the ingenious. And nowhere more than in reconnaissance satellites has ingenuity – 'American know-how,' if you will – found fuller play.

Unfortunately, the men and companies who design, build, and run these miraculous devices have never been allowed to take credit for their work. In fact, for sixteen years the United States government officially pretended that their work didn't even exist. Until mid-1961 the Defense Department freely admitted that it was developing satellites to reconnoiter foreign territory and to warn of missile attack – it even published budget figures for those programs. Then the Kennedy Administration ordered complete secrecy on the subject. Not until President Carter made a general reference to photographic reconnaissance in a 1978 speech did the U.S. government ever again admit the existence of such satellites.

It's not that the Soviets didn't know our satellites were spying on them. At first they complained, and they even proposed a United Nations ban on reconnaissance from space. By the end of 1963, though, they were no longer making such proposals: they had started launching their own recon satellites in April 1962. The reluctance of either side to acknowledge the existence of spy satellites clearly was not aimed at preventing the other from knowing about them. What it may have been aimed at was a mutually useful face-saving. If we didn't rub the Russians' noses in the fact that we were photographing their territory daily, they might feel less pressure to stop it by trying to shoot our satellites down. If we tacitly agreed to let their satellites scan us freely, they might not object to our doing it to them. In 1972 the two Superpowers even formalized their little agreement by including in their treaty limiting anti-ballistic missile systems an undertaking not to interfere with 'national technical means of verification' – their joint euphemism for spy satellites.

With the secrets of these satellites so jealously guarded, it is difficult for those without security clearances beyond 'Top Secret,' without the officially sanctioned 'need to know,' to find out anything at all about them. Several kinds of clues can lead to fairly firm inferences about both American and Soviet reconnaissance satellites. First, the orbital characteristics of all earth satellites are, sooner or later, public information. Countries launching satellites register them with a United Nations bureau. NASA's Goddard Space Flight Center publishes a bimonthly 'Satellite Situation Report' listing all the objects in orbit that are tracked by the Aerospace Defense Command. (The NASA report sometimes omits the current information on some U.S. military satellites, but the information is usually available elsewhere, if a little later.) The British Royal

Aircraft Establishment publishes another satellite log based on its own tracking information.

The 'orbital elements' – the measurable characteristics of a satellite orbit – reveal a good deal about the probable purpose of a satellite. *Apogee* and *perigee* are the points at which the satellite is farthest from and nearest to the surface of the earth. The greater the difference between the two, the more 'eccentric,' or elliptical, is the orbit; if the altitude is the same at all points – if the apogee and perigee are essentially identical – then the satellite is in a circular orbit. Particular kinds of orbits are more suitable to particular kinds of missions. For example, just about the only kinds of satellites to fly with perigees under 100 miles are photographic reconnaissance satellites, swooping in for a very close look at what's below.

An orbital element that depends on the apogee and perigee is the *period*, the time it takes for the satellite to make one circuit around the earth. As we saw with communications satellites, particular periods can be utilized for particular purposes. U.S. photographic reconnaissance satellites and some weather satellites often follow what are known as 'sun-synchronous' orbits: each pass of the satellite over a particular region on the earth's surface is at about the same local time of day, giving the photo interpreters or the weathermen a series of comparable views of the same scenes.

The *inclination* of a satellite's orbit – the angle it makes with the equator on its northerly pass – can also give clues to its function. For example, a low-flying satellite in a highly inclined orbit – one that passes near the poles – will eventually pass over almost all geographic regions: it's a good candidate for being a reconnaissance satellite, especially if its altitude is relatively low. (Weather and navigation satellites also often travel high-inclination orbits, but usually at somewhat higher altitudes.)

A piece of orbital information that's easily available is the lifetime – the length of time between the date of launch and the date of 'decay' ('decay' is the term for the re-entry of a satellite into the earth's atmosphere, whether the return is from natural causes or by human command). As a rule, the United States and the Soviet Union either order or allow their reconnaissance satellites to decay when they're no longer usable. The resulting lifetime tends to be characteristic of particular kinds of spacecraft.

Another useful piece of information about satellite orbits is called 'right ascension' – a measure that locates the plane of the orbit with respect to the stars, and therefore allows prediction of the tracks the satellite will follow over the rotating surface of the earth (which itself is circling the sun). Sophisticated private satellite watchers (particularly members of the British Inter-planetary Society) can and have analyzed the orbital planes of U.S. and Soviet satellites to make deductions about their uses. This kind of sleuthing can be tricky because some of those satellites carry maneuvering rockets to change their orbits or to maintain them against the natural forces of change.

If the earth were a perfectly isolated sphere in a perfect vacuum, and if a satellite around it were the same, the orbital elements would never change, they would be mathematically predictable forever. In fact, several forces act to alter orbits from the ideal. At lower altitudes – where photographic reconnaissance satellites operate, for example – there's still enough thin atmosphere to resist the hurtling object, to gradually rob the satellite of its momentum, causing it to sink ever lower and eventually to burn up in the denser air it finally encounters. And even satellites traveling high enough not to decay for many hundreds of years are still subject to various perturbing forces that make them vary from Kepler's and Newton's laws.

One such force is the stream of energy and particles flowing from the sun, especially when they are caught up in the earth's magnetic field. Another is the combination of varying gravitational pulls from earth, sun, and moon as they revolve around one another. Still another is the imperfect shape and density of the earth itself, which makes the gravitational field around it somewhat irregular – and so does the same to the orbits of the earth's satellites. Sometimes, though, these perturbations are actually useful. Sun-synchronous orbits, whose planes are lined up with the sun, wouldn't be expected to stay that way as the earth-satellite combination revolves around the sun during the year, but the satellite launchers can plan the orbit so that it perturbs – rotates its plane – in synchronization with the earth's circuit of the sun.

So by studying the orbital elements of satellites we can make some deductions about their purposes. But what about their capabilities? Can a spy satellite really read a book over the shoulder of a man on the ground? A newspaper headline? The license plate on a car? Without direct access to inside information, we can still draw some reasonable conclusions as to what spy satellites can and can't do. One way to derive those conclusions is to look at what's physically possible, given the laws of nature, and at what seems to be technically feasible, given the publicly known 'state of the art' in the field of 'remote sensing.'

Over the years the CIA and the Air Force have developed several ways of taking pictures from space. In the first system – the kind Discoverer 13 carried – a telescopic camera took the pictures, the satellite mechanism (built by General Electric) wound the exposed film onto a reel in a detachable capsule, and the capsule was ejected earthward. When friction with the atmosphere had slowed the capsule enough, its parachute popped out and it floated on down. Special aircraft of what is now the Air Force 6594th Test Group at Hickam Air Force Base, Hawaii, flew out to meet the descending treasure, trailing a trapeze that snatched the parachute in midair. (Sometimes they missed: then a helicopter had to fish the capsule out of the Pacific.) The film next went to Washington for intelligence interpretation. The Soviet Union started using a similar method in 1962, except that they apparently recovered the entire satellite, not just a smaller nose cone. In fact, to this day the film-return method is still the one the Soviets use predominantly – between 1962 and 1980 they had launched more than 500 such satellites.

By 1962, though, the United States had worked out a way to get pictures back more directly from a satellite. The camera (developed by Eastman Kodak) not only took the pictures, but developed the film right on board the spacecraft; then a TV scanner (from CBS Laboratories) scanned the film and transmitted the pictures down to Air Force earth stations. We know how the system worked because NASA later used a quite similar device on the Lunar Orbiters to survey landing sites for the Apollo missions. These satellites would go into orbit for three or four weeks and send back wide-area surveys of the target area. Then the film-return satellites would be sent up for just a few days, returning more detailed views of the interesting features the photo interpreters might have found in the area surveys.

How close a look can such a satellite take? The usual standard of photographic closeness is 'ground resolution,' the size of the smallest object one can distinguish in the picture. The resolution of a satellite picture depends on many factors: the wavelength of the light recording the image, the fineness of the film (or whatever other sensitive surface is at the focal plane of the camera), the quality of the optical system, the focal length of the telescope, the aperture of the telescopic lenses or mirrors (which determines how much light the telescope can gather), the altitude of the satellite (how far away the object is), the haziness of the atmosphere, and the amount of contrast between the object and its background.

Let's just assume for the moment that the best telescope on a photo reconnaissance satellite has a focal length of about 20 feet (the satellite doesn't have to be that long: the optical system uses 'folded' optics – a combination of mirrors that bounces the light around corners for a total path length of 20 feet). Suppose further that the wavelength of the light is in the visible range and that the aperture and film grain give a combined film resolution of about 175 (separately discernible) lines per millimeter. Such a camera aboard a satellite at an altitude of about 90 miles should, in theory, be able to distinguish an object on the ground about 5 to 6 inches across.

So, by looking at the basic physics and by making reasonable assumptions about satellite camera technology, we can come up with a plausible ground resolution for the better film-return spy satellites. But there's yet a third set of clues to what is possible in space reconnaissance: the press leak. In Washington, secrecy ('classification for national security purposes') and leaks are two sides of the same coin: the control of information. Higher officialdom likes to control what the bureaucrat, Congress, and the public know and when they know it. Sometimes a President or a Secretary of Defense will decide openly to 'declassify' a piece of information – as when Jimmy Carter finally acknowledged that the U.S. does in fact operate photo reconnaissance satellites.

Then, of course, it's not a leak, because by definition the information is no longer classified. At other times, though, a President or one of his staff, or a Cabinet member, may call in a reporter and give him a piece of inside information 'off the record,' but expecting to see the story on the next day's front page. In that case, the leaking public official wants to strengthen his political position on some issue by getting out his side of the story without appearing to break the rules of classification.

In other cases, though, lower officials may leak a story to the press because they believe their own position is losing out in the internal policy forum. That's the kind of leak the higher officials usually like to plug. In 1981, for example, the Joint Chiefs of Staff made the case in a secret meeting that even as much as the Reagan Administration was planning to spend on the military in the next five years (about $1.5 trillion), it was, in their view, still about $750 billion short. Someone with access to that information apparently decided that public disclosure of the Chiefs' view might help build political pressure for more defense money. What the leak did lead to was an Administration crackdown on leaks, complete with lie-detector tests and insistence on advance approval of all press interviews dealing with 'classified matters.'

One journal that editorialized against the crackdown was *Aviation Week and Space Technology*. *Aviation Week* might be called '*Aviation Leak*,' so frequently do its pages bear information that is 'classified.' *Aviation Week* is a trade journal to the aerospace industry and, hardly surprisingly, is a vigorous advocate for more and better military technology. So those in the Pentagon and in industry who want to build support for their side of any particular story often find *Aviation Week* reporters to be sympathetic listeners. And to maintain the symbiosis, those inside 'sources' feed the journal a continuing stream of inside information, even when it may not bear on a current controversy. The result is that *Aviation Week* is a useful 'open' source of information about military technology.

Over the years, the magazine has published a good many details about reconnaissance satellites. When one of its articles says that the ground resolution of the Air Force's high-resolution, film-return satellite is about 6 inches – when that piece of information is consistent with the publicly available orbital elements of the satellite the article identifies, and when it jibes with what we can calculate by making reasonable assumptions about the camera technology – then the information is probably pretty reliable.

What is the meaning of that 6 inches of ground resolution to the photo interpreters? Some official testimony in Congress gives a fairly good idea. Just to detect the existence of a ballistic or air-defense missile on the ground requires a resolution of 10 feet; to identify the missile in a very general way requires a resolution of 5 feet; if the interpreter has a set of known types of missile within a general class, he can discriminate the particular type he is looking at with a resolution of 2 feet; a more exact description of the missile site takes a resolution of 1 foot. It takes only 15 feet of resolution to spot an airplane on the ground, but it takes 6 inches to describe the exact dimensions and outfitting of the plane.

The common rumors about what the photo interpreters can do probably don't apply. Reading the print in a book from a satellite is just beyond the theoretically possible, which is around 2 to 4 inches of ground resolution. But even that (and the very best telescopes and cameras under the very best lighting conditions may well have approached the theoretical limits) isn't good enough to make out newspaper headlines or automobile license plates.

The highest-resolution satellites have been used in conjunction with somewhat higher-flying vehicles that do the area surveillance, sweeping over much wider swaths of territory and gathering the information available at lower resolutions as well as giving the interpreters clues as to likely targets for more detailed examination. In the 1960s, the satellites that developed their film on board and sent it back by television did this. These satellites could operate for a month, sometimes longer, before running out of film and fuel. After interpretation of the 'area surveillance' TV pictures, the Air Force could send up the higher-resolution cameras for closer looks.

In 1971, the U.S. launched the first of what became known as the Big Bird series of reconnaissance satellites. The Big Bird rides into a polar orbit on top of a big Titan III-D rocket: the satellite and its built-in maneuvering rocket probably weigh close to 29,000 pounds. The Big Bird can send area surveillance TV pictures back for about six months (the longest any Soviet reconnaissance satellites have yet been kept in orbit has been about a month). But it also sends back film capsules (*Aviation Week* says four separate ones) with more detailed pictures of selected targets – so combining in one satellite what used to require two.

With a perigee of about 100 miles, the Big Bird flies a little higher than the highest-resolution cameras, so the Air Force still sends the close-look type up about once a year. These 3.5-ton vehicles stay up for fifty to eighty days and send back at least two film pods. One went up on February 28, 1981, when American concerns about a possible Soviet military intervention in Poland were rising. Six weeks later, at a NATO meeting in Bonn, Secretary of Defense Weinberger reportedly displayed satellite pictures of Soviet troops in Poland and of the latest in Soviet weaponry.

As useful as these pictures are, they are just one of several kinds of reconnaissance image: other images, even at much lower ground resolutions, can yield information at least as valuable as the dimensions of some particular weapon. Infrared photography (in wavelengths too long to be seen by the human eye) has long been used in aerial and space reconnaissance. The magazine *Defense Electronics* recently reported that the U.S. has a 'major sensor camera that optimizes heat sensitivity in order to photograph images in the past. The purpose of the camera, kept under wraps for 25 years, is to detect missile launching sites and troop movements that were completed in the past and camouflaged today.' The newest satellite imaging devices make use of several bands of infrared light: they have sensors to register wavelengths that photographic films can't see. A satellite that almost certainly carries these 'multispectral scanning' devices is the CIA KH-11.

William Kampiles, at twenty-three, was a junior officer in the CIA Operations Center, the Agency's screening center for incoming intelligence. Among the reference material in the Operations Center was the manual explaining the kinds of information available from the KH-11 satellite. Some people did (and some still do) confuse the KH-11 with the Big Bird: they both weigh more than 14 tons, and they are both probably about 10 feet in diameter and 40 to 50 feet long. But a careful observer of the satellite logs would have noticed that while the Big Bird flies as low as 100 miles, the newer satellite (first launched December 19, 1978) flies at perigees between 150 and 180 miles. And while the Big Bird stays on orbit for about six months, the KH-11 flies and transmits data for more than two years. The Soviets must have watched with great interest as the KH-11 crossed their skies. And they must have been delighted when the young Kampiles approached their embassy in Athens in March 1978 and offered to sell them the KH-11 for a mere $3,000.

Thanks to Kampiles, the KGB knows more about what the KH-11 can do than you or I are allowed to. Still, we have some clues and some information on this big spying machine. Earlier wide-area surveillance satellites recorded images on film, developed the film, and transmitted to earth a television picture of the developed images. In contrast, KH-11 telescopes scan the territory below with light-sensitive electronic sensors, and those sensors translate the light patterns into digital signals (numbers for radio transmission). The sensors are 'filtered' to detect certain bands of visible and infrared light (including infrared bands that no film can record). The Skylab manned space station carried a 'multispectral scanner' something like this, and so does the Landsat 3 earth resources satellite. But because the KH-11 flies at lower altitudes than those satellites, and because it almost certainly has more powerful telescopes, it's likely to produce scenes with much higher ground resolution than the Landsat pictures we see in magazine ads.

Infrared images reveal far more to the intelligence analysts than do pictures taken in visible light. Thermal infrared, or 'heat,' is emitted rather than reflected, so the scanners can produce pictures at night. Objects of different temperature emit different infrared 'signatures,' and objects of different surface quality reflect light differently, even though they may appear to be the same color. Therefore the analysts can use the multispectrally scanned images to detect military camouflage by differentiation between, say, live and dead or artificial and real foliage. Hidden objects will probably hold heat differently from their background environment: their infrared signatures may give them away.

We know how Landsat images help agricultural scientists detect the extent, variety, and health of crops; surely the CIA watches Soviet agriculture carefully. But the multispectral images must also be useful in studying Soviet industrial processes. Geological surveys from space tell what minerals are mined where. Observation of the temperatures of particular parts of chemical and manufacturing plants would reveal the kinds of processes going on in the plants and what their output is. (In discussing their estimate of Soviet military production, CIA analysts openly admit that they rely in part on their knowledge of the square-footage of factory space the Soviets use to produce each type of weapon.)

The satellite-returned images are all the more meaningful because the U.S. intelligence agencies have a wealth of historic photographic records with which to compare them. The analysts have been able to watch the Soviets build their factories, dig their missile silos, construct their military bases, send their ships to sea, set up their encampments, for twenty years. In 1981, for example, the Pentagon published a booklet on Soviet military power that showed an 'artist's conception,' in considerable detail, of an SS-20 mobile ballistic missile launch site. The same booklet laid out the dimensions of the largest Soviet tank plant.

In Congressional testimony in 1981, Defense Intelligence Agency officials reported that 'there are 134 major final assembly plants involved in producing Soviet weapons and end products. In addition, we have identified over 3,500 individual installations that provide support to these final assembly plants.' The officials said that 37 plants produce aircraft material and 49 produce missile material. They also provided a table showing production numbers for several types of Soviet weapon.

Comparisons over time are one way of wringing more information out of the images from space. Computer enhancement of the images is another. With techniques developed in the past ten or fifteen years, the photo interpreters can perform magic on the raw images they get from satellites: they break each picture into little dots called 'pixels,' then manipulate the pixels again and again. Each pixel has a numerical brightness index, so the contrast in each part of the picture can be varied to bring out subtle contours and detail. The analysts can combine the pixels from two or more pictures of the same scene to produce new pictures, more detailed than any of the single shots.

Overlaying the multispectral images, assigning 'false colors' to the hues and tones latent in the images, they can reveal new patterns in a scene. Using sophisticated computer programs to 'fill in the dots' between pixels, the analysts can sharpen the lines of resolution beyond the theoretical limits of the optical systems that took the pictures. We've had just a taste of the possibilities of digital image enhancement in the pictures of Jupiter and Saturn that the Jet Propulsion Laboratory put together from the digital transmissions of the Voyager spacecraft.

Satellite reconnaissance is obviously very useful in estimating what the Soviet military can do to us and our allies. At the same time, those pictures have directly shaped modern U.S. nuclear strategy. The continuing evolution of more and more accurate nuclear delivery systems, coupled with theories of 'counterforce' strikes (aimed at Soviet military installations, especially missile silos) and theories of 'limited' nuclear war against key military and political targets, simply would not have been possible without precise satellite mapping of Soviet territory and potential nuclear aimpoints.

Not only do the satellite photos yield the exact locations of the targets, but they can also tell the missiles and bombers how to get there. In the past few years, the Defense Mapping Agency has added new employees and equipment solely to convert satellite pictures into three-dimensional 'digital maps' of the Soviet Union. The new U.S. air-launched, ground-launched, and sea-launched missiles will store portions of those maps in their own computers. Using radar altimeters to compare the terrain below with the stored maps, the missiles navigate themselves to within 100 yards of their targets deep inside the Soviet Union.

As useful as the imaging reconnaissance satellites have been, they're not the only means of gathering military intelligence from space. Other kinds of sensors exploit other bands of the electromagnetic spectrum. The class of intelligence commonly known as ELINT (Electronics Intelligence) includes the interception and analysis of foreign communications and radar signals. The first U.S. ELINT satellite, a 1-ton spacecraft, went up in March 1962. About a year later, that satellite was replaced by a 3,300-pound vehicle that went into nearly polar orbits at altitudes of around 300 miles (a later version, first appearing in 1968, approached 4,400 pounds).

By 1963 the U.S. had also developed a 120-pound ELINT satellite, sometimes referred to as a 'ferret,' that 'hitchhiked' into orbit with photo reconnaissance satellites – first with the 'close-look' satellites and then, beginning in 1966, with the area surveillance type. This little hitchhiker would use a smaller rocket to pop into a slightly higher orbit (around 300 miles) than its big companion. The last king-size ELINT satellite went up in 1970. The hitchhikers started riding with the Big Bird when it went into service in 1971. Since they now do so only occasionally, it's entirely possible that the Big Bird and the KH-11 have taken over some of the electronic eavesdropping work of the older 2.5-ton ELINT satellites. The Soviet Union continues to send up three to five large (1 ton and 4 ton) ELINT spacecraft a year.

What kinds of information come out of these missions? Radar signals reveal how air defense systems work, and also give the electronics countermeasures ideas on how to evade or jam the enemy radars. The Soviets, for example, have about 10,000 anti-aircraft missiles, using radar for finding their targets, that U.S. bombers would have to get past in a nuclear war: the U.S. B-52s and FB-111s carry complex jamming devices to help them evade detection and destruction. Intercepted radio communications may or may not reveal important details about the immediate plans of enemy military forces, but they can give clues to the routines and procedures of day-to-day military operations, and help the analysts fill in their picture of the enemy 'order of battle' – the composition, location, and readiness of the other side's armed forces.

Satellites are only one of the ways ELINT is gathered. During the Tonkin Gulf incident of 1964, the U.S.S. *Maddox* was cruising off the North Vietnamese shores with technicians and special equipment from the National Security Agency, the main U.S. collector of electronic intelligence. The ship's mission was to stimulate responses from North Vietnamese coastal radars, in preparation for possible future air or sea attacks. The U.S.S. *Pueblo*, captured by the North Koreans in 1968, was also collecting ELINT. Electronic reconnaissance airplanes skirt the borders of the Soviet Union on similar missions. And the U.S. still maintains some ground bases as 'electronic listening posts.'

The United States used to maintain an electronic listening post in Iran to intercept the telemetry from Soviet rocket launches, particularly its missile tests. Instruments aboard the rockets report to the ground their measurements of acceleration, position, fuel consumption, and other performance characteristics. Analyzing the intercepted telemetry tells the CIA and military analysts what the

Soviet missiles can do (and, in past years, helped verify compliance with strategic arms limitation agreements). The Soviet Union has an easy time monitoring U.S. tests: the missiles take off either from submarines in international waters or from Vandenberg Air Force Base, right on the West Coast. The Russians just pull their ELINT ships right up next to the launch sites and listen away. But they conduct most of their own ICBM tests at the Baikonur Cosmodrome (a long way from Baikonur), near Tyuratam, in Kazakhstan, in the south-central Soviet Union. The U.S. has listening posts and tracking radars in Turkey that help some in watching the Tyuratam tests, but distance and intervening mountains make the location less than ideal. Iran was ideal, but the two sites there of course fell after the Shah did. Rumor has it that the U.S. now has a listening post in China. But it also has one in space.

We learned about the KH-11 from one spy trial – that of William Kampiles. The existence of the Rhyolite series of ELINT satellites surfaced with the case of the Falcon and the Snowman (writer Robert Lindsey's nicknames for spies Christopher Boyce and William Lee). The CIA has had the Air Force launch at least four space vehicles that at first might be mistaken for missile early warning satellites: they're in geostationary orbit over the Indian Ocean. They're also more or less due south of the chief Soviet ICBM test site in Kazakhstan.

It came out during the Boyce-Lee trial that data from these satellites goes to a readout station in Australia – but *not* the one that monitors the early warning satellites. Instead, the data goes to a CIA-run complex in a valley called Pine Gap. After the Pine Gap computers have processed the data, it goes to the TRW Defense and Space Systems headquarters in Redondo Beach, California. That's where Boyce picked up the information on the Rhyolite satellite (only one was up when he did his spying) that his friend Lee sold to the Russians.

The Rhyolite series (it may have a different name by now) is apparently a general collector of electronic intelligence, but its most important role is to intercept the telemetry from those Soviet ICBM tests. Sorting out the

VHF missile telemetry signals from all the other VHF broadcasts in the satellite's view must be an enormous task, and squeezing useful information from the other broadcasts as well must also tax the Pine Gap computers. Reports about the Boyce-Lee trial suggested that the Rhyolite data went from Space Park in Redondo Beach to CIA headquarters in Virginia.

But a more likely destination for Rhyolite and other missile-related data is Fort Meade, Maryland. Fort Meade is the site of the National Security Agency, which seems to carry out most ELINT operations. Also at Fort Meade is an even more obscure intelligence organization: the Defense Special Missile and Astronautics Center, or DEFSMAC, headed by a colonel from the Defense Intelligence Agency. This is a likely place for the computer number-crunching needed to make sense out of ELINT data. (Although the CIA employs about 1,200 image interpreters, much of the analysis of satellite-returned images probably goes on at the National Photographic Interpretation Center at the Washington Navy Yard.)*

The imaging and electronic reconnaissance satellites are invaluable for drawing up U.S. battle plans and keeping track of the other side's 'order of battle.' By keeping an eye on the potential enemy's economy and industrial plant, his weapons tests, and the large-scale disposition of his forces, the satellites can help our military forecast his 'intentions and capabilities.' Such arms control agreements as we've had with the Soviets probably wouldn't have been possible without the spaceborne 'national technical means of verification.' But what most of those satellites have *not* done is satisfy military desires for 'real time' information on Soviet military activities. 'Real time,' a term military

* Hidden within Air Force Intelligence is the National Reconnaissance Office. The NRO operates most of the intelligence satellites for the 'intelligence community' – which includes the CIA, the DIA, the NSA, and the service intelligence branches. The National Executive Committee for Reconnaissance, made up of the Assistant Secretary of Defense for Intelligence, the Director of Central Intelligence, and the President's National Security Adviser, rules the NRO budget. A committee of the U.S. Intelligence Board (chaired by the CIA director and composed of the heads of the various intelligence agencies) decides what targets the satellites (and spy planes) should go after. The NRO draws up a Joint Reconnaissance Schedule for all the hardware.

jargon has borrowed from computer jargon, means information about what's happening now, or near enough to now to permit a timely reaction.

There are inevitable delays, for example, in getting useful information from the TV pictures from Big Bird and the digitally encoded images from the KH-11. For either one, the satellite controllers have to wait until it passes over a ground station (in most cases one of the Satellite Control Facility's Remote Tracking Stations) to call down the images stored up during the satellite's pass over the Soviet Union (or whatever other territory is of interest at the moment). A satellite at altitudes between 100 and 250 miles is within line of sight of a ground station for only about six or eight minutes at a time (and then it has to be some degrees above the horizon before usable radio transmissions are possible).

One of the Remote Tracking Stations – the one near Thule, Greenland – is so close to the North Pole that it is, fortunately, frequently in the line of sight of the KH-11 and the Big Bird immediately after their passes over Soviet territory. So important is this station that it maintains a special link with the Sunnyvale headquarters (the Big Blue Cube) via the Satellite Data System communications satellites. The link is at a high enough frequency (1,760 to 3,000 megahertz) to transmit up to 256,000 bits of data per second, no doubt much of it originating with the imaging reconnaissance satellites.*

Should some of the analyzed images prove especially interesting, the satellite controllers can command the satellite to take a closer look on its next pass over the same area. Since these satellites take about ninety minutes to go around the earth once, and since the earth in the meanwhile revolved another 22.5 degrees, the satellite will not pass within line of sight of the same ground target for another sixteen times around. (The satellites are somewhat maneuverable, so the periods between passes over particular points can be altered a little, but not much. What the

---

* It may also be – although the Air Force will not confirm it – that the KH-11 beams its data directly up to the SDS satellites, making the images even more like 'real time' pictures.

maneuvering does allow is for the controllers to plan passes over desired areas at designated future times.)

So getting information from today's imaging and electronic reconnaissance satellites is a time-consuming, if worthwhile, set of tasks. One type of satellite that does provide 'real time' information goes by the uninformative name of Defense Support Program, or DSP. The DSP satellites, one watching the Soviet Union from over the Indian Ocean, the other two watching the Atlantic and Pacific oceans from their geosynchronous stations over the Western Hemisphere, warn of missile launches from land or sea.

Each DSP satellite (TRW builds them) is about 22 feet long, with a 9-by-9-foot solar-cell-covered cylinder for the electronics at one end and a 12-foot-long telescope at the other. In the focal plane of the telescope is an array of 2,000 detectors, sensitive to the infrared radiation of missile exhaust plumes (the satellite also has ultraviolet sensors and a televisionlike system, but the infrared detectors are the most precise). The telescope is offset by about 7.5 degrees from the axis of the body of the satellite; the whole satellite spins at about five to seven revolutions per minute. This means that the telescope as it spins scans large regions of the earth's surface. At any given instant, each of the detector cells covers about 1.2 square miles. As the telescope scans, its detectors register a count of events – missile plumes detected – so the earth stations (in Australia and Colorado) that process the data can roughly deduce the number of missiles launched. So far, of course, the number has been one or two at a time, in missile tests and space launches.

The Air Force has had TRW 'upgrade' the most recently built DSP satellites. One improvement is the addition of detectors sensitive to a second, longer wavelength of infrared light, to make it more difficult for a Soviet ground-based laser to blind the system. And where the old version had 2,000 detector cells in the focal plane, the new one has 80,000, yielding much more precise information about where the rockets are launched from and where they are going. The other changes are intended to improve the 'survivability' of the satellite system in a nuclear war. These include the ability to store and rebroadcast the data the satellite gathers and a satellite-to-satellite communications 'crosslink.' Today, the satellite in the Eastern Hemisphere sends its data down to the ground processing station in Australia, that station then sends the processed data up to a Defense Satellite Communications System satellite, which in turn sends the data to the U.S. command posts. With the crosslink, the Eastern satellite will be able to transmit its data via the Western satellites directly to a ground processing station inside the United States.

The upgraded satellites will also do more of the initial computer processing of the data themselves, making it possible to deploy mobile 'simplified processing stations' in the U.S., in addition to the large, fixed station now at Buckley Field outside Denver: the idea is that the mobile stations are more likely to survive the initial wave of a nuclear attack. And since the Satellite Control Facility ground stations are also vulnerable to nuclear attack, the newer DSP satellites will also be more autonomous: they'll be able to 'keep station' longer without commands from the ground. The first of these upgraded satellites will probably go up no later than 1984.

The DSP satellite also carries radiation sensors to detect nuclear explosions. The United States used to launch a separate series of TRW-built satellites, called Vela Hotel, for that purpose. The Vela Hotel ('Vela' is the general designator for nuclear explosion detection programs, whether space-based or earth-based) satellites went into 70,000-mile-high orbits. For full earth coverage they were launched in pairs. The last pair, Vela Hotel Eleven and Twelve, were launched in 1970; ten years later they were still working. The next generation of nuclear explosion detectors will be riding on the new Global Positioning System navigation satellites.

There is one more set of satellites that provides 'real time' military intelligence: the Navy's ocean reconnaissance satellites. The U.S. ocean reconnaissance satellites go up in a package into a circular orbit about 600 miles high and at an inclination of 63.5 degrees to the equator. Then two or three sub-satellites split off from the mother ship, separating in orbital plane so that the satellites can observe overlapping swaths of sea below. Radio and microwave receivers on the satellites pick up the radar and communications signals (and, possibly, the reflections of natural microwaves) from the ships below. Computer comparison of the signals lets Naval Intelligence analysts vector in on the locations of the Soviet vessels.

The Soviet Union operates similar ocean reconnaissance satellites, and another kind as well. They have a satellite with an active radar system, powered by a nuclear reactor. (The most famous of this series was Cosmos 954, which crashed into the Canadian Northwest Territories on January 24, 1978. Normally, if such a satellite is losing altitude or has completed its mission, the Soviets send the reactor section into a higher orbit, where it's unlikely to decay and strew its radioactive core across the countryside. That time, the mechanism failed, and Canada, with considerable U.S. help, ended up with a major search and cleanup job, which they dubbed Operation Morning Light.) Such a powerful radar system yields much more precise and timely information. During the British-Argentine war over the Falkland Islands, the Soviets put up fresh copies of both types of ocean reconnaissance satellite. There was some speculation that they may have been passing on some of the information from those spacecraft to the Argentines.

U.S. military worry that in war the Soviets would use their ocean reconnaissance satellites to pinpoint targets for their long-range Backfire bombers and anti-ship missiles. U.S. officials have in fact announced that those satellites would be the Number One target for a U.S. anti-satellite weapon.

The Navy has for some time been interested in the idea of a radar-scanning reconnaissance satellite. For some years, it carried on a development project to that end known as Clipper Bow. Congress eventually forced cancellation of Clipper Bow, but the Navy is trying again. The latest proposal is for an Integrated Tactical Surveillance System (ITSS). One element of the ITSS satellite is likely to be active radar, which can see through darkness, clouds, and rain. The Navy wants the satellite to spot those Backfire bombers as well as ships. And it wants the data from the satellites to be refined enough that Navy ships can use it without waiting for big on-shore computers to process it.

The Navy ITSS project is just one sign of a growing U.S. military demand for 'real time' satellite reconnaissance. The Air Force Space Division and the Electronic Systems Division, for example, have been commissioning contract studies on an even larger space-based radar systems with companies like Lockheed, Grumman, and General Dynamics. They've been considering orbiting antennae 130, 300, even 600 feet in diameter (delivered by the shuttle), possibly at geosynchronous altitude, possibly much lower. Such giant radars could track ships, air planes, cruise missiles, maybe even tanks, and do so through any weather.

What makes these military dreams of worldwide, 'real time' reconnaissance and surveillance possible is the miracle of micro-electronics. The integrated chip microcircuit has of course already transformed the way business, government, the military, and consumers do things (I'm writing this book on a desktop computer; on my wrist is a device that's a watch, an alarm clock, a stop-watch, a calculator, *and* a 'space invaders' game). Contrasted with the vacuum tubes of thirty years ago or even the transistors of twenty years ago, today's microchips are astonishing. But progress in electronic miniaturization and speed is just beginning.

The terms to watch are 'VLSI' and 'VHSIC.' Very Large Scale Integration (VLSI) is what we've seen so far and what we'll see more of: the cramming of ever more electronic circuitry into ever smaller spaces. Very High Speed Integrated Circuits (VHSIC) are the next step: circuits that are not only smaller but that process data faster than ever before. In the past few years the Army, the Navy, and the Air Force have been pumping several extra millions of dollars into universities (Stanford, Cornell, UCLA, Illinois, Carnegie-Mellon, and others) and companies (General Electric, Honeywell, Hughes, IBM, Raytheon, Rockwell, Texas Instruments, TRW, Westinghouse are the main ones) doing research in VHSIC.

The individual elements on today's 'state of the art' micro-circuit chips are perhaps 5 to 7 microns (millionths of a meter) wide – several times thinner than a human hair. The features on a new IBM chip are 2 microns wide; Honeywell has made a chip with 1.25-micron-wide features; and an experimental TRW chip has 1-micron elements. The goal of the Defense Department VHSIC program is microcircuits with features only .5 micron wide. (As the editors of *Aviation Week* point out, if one were to print a map of the United States on a sheet of paper only 20 inches wide, and the lines of the map were only .5 micron wide, that map could show every single street in the country.)

This kind of performance is especially important to space-based reconnaissance and warning systems. For example, the focal plane of the telescope in the original version of the DSP missile warning satellite had about 2,000 detectors wired together in an array that scanned across a large area of the earth's surface, ready to make a count of missile events as they swept by the rotating telescope lens. It's now possible to join much larger numbers of detectors into what are called 'monolithic arrays' – single printed chips without wires between the elements. The new detectors, called 'charged-coupled devices' (CCD), translate the incoming light directly into electrical impulses, and the detectors work together to produce an 'integrated' signal out of the impulses. Honeywell has built an infrared focal-plane array for the Air Force that contains 80,000 detector cells. The Itek Corporation (a long-time manufacturer of reconnaissance equipment) advertises an electrooptical focal plane 'with special silicon CCD imaging microcircuits containing a 2,048-by-96 [i.e., 196,608]-element array of 0.0005-inch detectors.' What's more, those arrays can be joined together in long strips to make even larger arrays.

In the case of the warning satellites, the new technology means that future telescopes will not have to scan over a broad area, but will be able to 'stare' at that area continuously. This staring has several advantages. It will be harder to jam the system with a laser on the ground, because each individual detector cell will stare only at one spot, not scan across a large area, so the laser would only affect a relatively few cells. (The new systems will also have 'tunable' spectral filters, allowing the sensors to work at several wavelengths and possibly avoid laser blinding at some particular wavelength.) Another advantage of the staring sensor with thousands of detectors is that the resolution of the optical system is much higher. The current generation of warning satellites reports that approximately so many missiles have been launched from such and such an area; the next generation will report exactly how many from exactly which silos (in the case of fixed, land-based missiles) and will be able to track the trajectories of those missiles farther, giving better predictions of where and when they will land (and sensitive new detectors may tune in on the infrared wavelengths not only of the missiles but of their separated multiple warheads, farther along in the trajectory).

The military planners want this kind of information for two reasons. First, if they decide to 'launch under attack' – to launch American ICBMs before the Soviet warheads arrive – the warning satellite data will tell them which silos are under attack and when they will be destroyed. Second, in a retaliatory nuclear strike against the Soviets, the U.S. military men want to aim their missiles at the Soviet silos that haven't fired their missiles yet; the new warning satellites will tell them which is which.

The Defense Advanced Research Projects Agency (DARPA) has outlined the design of a satellite it plans to launch on the space shuttle in 1988 called the Advanced Sensor Demonstration. This Advanced Sensor will go into geosynchronous orbit like the current warning satellites. But it will detect and track not only missiles but strategic bombers and tactical fighter planes as well. According to a DARPA report, '. . . the preliminary configuration of the sensor payload incorporates a telescope with optical filters, a focal plane, a cryogenic refrigerator to cool the focal plane [so that it can detect long-wavelength "thermal infrared" radiation], and a processor that controls the system and converts focal-plane data into target track information for use by small ground terminals.' The focal plane of the telescope will probably consist of numerous monolithic arrays joined together to build a mosaic of 1 *million* detectors. Although this satellite will just be 'experimental,' it will be available to the U.S. military '. . . . for observation of critical events anywhere in the world during most of the system's life [two and a half years].'

The super-miniaturization of focal-plane sensors is a necessary, but not sufficient, step in the coming progress in observation from space. As the term 'Very High Speed Integrated Circuits' implies, the new technology offers not just compactness, but speed. According to the director of the Westinghouse VHSIC program, the research may lead to a signal processor 'capable of performing 40 million complex-number operations per second on only two six-by-eight-inch printed circuit boards.' That's fifty times smaller than comparable present devices, and almost as fast as the biggest, fastest computers in the world today. The problem with trying to monitor military activities from space is not that the sensors will 'see' too little but that they will see too much.

From a torrent of electromagnetic emissions and reflections, the sensors, computers, and human users have to pick out the particular signals that carry the information they're looking for. The observation system, in other words, has to pick out the targets from the background clutter. Today's reconnaissance and warning satellites, like the KH-11 and the DSP, transmit the data their sensors collect in relatively 'raw' form to ground stations for computer 'signal processing.' The trend, though, is toward more 'on-board' processing, made possible by the advances in VHSIC. DARPA and the Air Force are developing an 'advanced on-board signal processor.' According to DARPA Director Robert Fossum, 'This processor was initially designed to process the raw data for space-based radar. As the design matured, it became clear from analyses that the design was general enough to encompass all of the known space signal process requirements foreseen through the year 2000.'

Fossum's agency is of course already laying the groundwork for the 1988 launch of its Advanced Sensor Demonstration. The first major experimental step was to fly a U-2 airplane on what were called HI-CAMP missions. HI-CAMP (High Resolution Calibrated Airborne Measurement Program) experiments used a prototype 'staring' infrared array to measure earth background and to identify the infrared 'signatures' of airplanes against those backgrounds. DARPA calls the next major experiment Teal Ruby.

The Teal Ruby sensor system (for which Rockwell is the chief contractor) is to ride on a satellite to be delivered to orbit by the space shuttle in 1986. (The satellite, which will carry other experiments as well, is part of the Air Force Space Test Program, and therefore carries the label 'P80-1,' identifying it as part of that series.) The satellite will go into an orbit with an altitude of 460 miles and an angle of inclination to the equator of 75 degrees. Teal Ruby will track 'cooperative' Air Force aircraft, but it will also take worldwide measurements of the infrared background against which it would have to pick out enemy bombers. Although its sensors are designed to make exact measurements of the 'signature' of the larger 'strategic' aircraft, the DARPA researchers also expect it to at least 'demonstrate the ability to detect and track much dimmer targets such as cruise missiles.'

Not surprisingly for advanced military technology, the Teal Ruby telescope has had more than a 100 percent cost overrun since Rockwell got the development contract in 1977. The original contract was for $25 million, with another $25 million anticipated for preparing the sensor for the actual space mission. By 1981, the sensor program was expected to cost $110 million – and was a couple of years behind the original schedule.

In the next decade not only these new 'warning' satellites, but the KH-11-like spy satellites will probably make use of the new large arrays of charged-coupled device detectors combined with on-board signal processors capable of prodigious feats of computation. An invention called the 'multispectral linear array' is likely to replace the scanners on the KH-11. The scanners use a mechanical mirror, sweeping left and right as the satellite moves ahead, to transmit light to an electro-optical focal plane with a few individual detector cells. This process produces relatively low resolution, suitable for large-area surveys but not for finer detail. And analysis of the data from the scanners requires a lot of processing by computers on the ground. The multispectral linear array will not scan from side to side, but rather 'stare' straight down at the territory below, something like a push broom sweeping across the ground track of the satellite, building a continuous scroll of electronically recorded images as it goes.

The detectors are more sensitive to lower intensities of light and infrared radiation than those in the current scanners, so better ground resolution should be possible. What's more, the signals need less processing than those from scanners; on-board processing could further reduce the need for length ground computations. With this technology, the reconnaissance satellites might be converted from longer-range 'strategic' use to 'real time' tactical use through mobile receiving stations in the field.

According to DARPA head Robert Fossum, the research at his agency and in U.S. companies promises 'revolutionary new options for the future defense of the country.' In particular, he said in 1979, 'Advanced forms of the sensor technology, where thousands of detectors are mosaicked on a single small silicon chip, is almost uniquely contained within the U.S. technology base. By applying this very powerful and cost-effective technology to opportunities in both tactical and strategic concepts, we can have the option for pulling significantly ahead of our adversaries in capabilities for surveillance, target acquisition, and homing guidance.'

This promise of 'pulling significantly ahead' of the Russians with a sort of technological jujitsu is surely tempting. Spacemen frequently apply the term 'force multiplier' to military space technology. What you may lack in numbers, you make up for in intelligence and agility. With a geosynchronous electrooptical observation post of great precision, with space-based radars, with a network of lower-altitude reconnaissance satellites – with all of these feeding 'real time' data to our military forces, we could have nearly divine omniscience. Our military commanders would always know where all of the enemy's ships, planes, and tanks were, and where they were headed. Combining that knowledge with the new precision-guided weapons on land, at sea, in the air, and in space, the commanders could attack the more numerous Soviet forces with vastly superior efficiency.

It's hard to argue against not just *more* bang for the buck, but more *effective* bang for the buck. It's one way to avoid drafting huge bodies of troops as the Soviets do. It plays to our own comparative advantage – technological ingenuity – in the arms race. But the 'technological leapfrog' also carries some risks that we need to be aware of. One of the risks is hubris: if we can survey the globe like gods, can we not also be omnipotent like gods? Armed with seemingly perfect 'survivable' command and control, with ultra-exact surveillance and warning, with absolutely precise weapons, will our leaders come to see nuclear war as a controllable process? The emerging doctrines of 'escalation control,' 'counterforce,' and 'protracted nuclear war' suggest tendencies in that direction. In particular, the notion of some of the spacemen that space is the way for the United States to regain its lost military superiority over the Soviets stems directly from the faith that a technological advantage in space could be decisive.

But if we do pursue that advantage, it doesn't necessarily mean that the Soviets have no possible responses. They may not be able to match our technological leverage (then again, they may: some suggest that the technological gap between us and them has been closing in recent years). To try to compensate for our advantage, they might, for example, try to saturate our defensive systems with ever more numerous offensive systems of their own. Then, even if we knew where all their weapons were and where they were headed, we might not have time and weapons enough to counter them. Or they might adopt strategies to try to neutralize our particular advantages. Take the case of the improved warning satellites which would tell us which

Soviet missile silos had launched missiles and which were still loaded: our ability to 'target' those silos would encourage the Soviets, if and when they use their missiles, to fire them all at once, lest they never get a chance to use them at all (so much for 'limited' or 'protracted' nuclear conflict).

Over time, we can also expect them to make progress in evening up the 'force multiplier' score with technical advances of their own. For example, it took them many years to master the MIRV (multiple independently targetable re-entry vehicle) multiple-warhead missile technology of the United States, but they did it; and leaked intelligence reports are that they have been able to make their missiles as accurate as ours.

As long as we are in an arms race with the Soviets, we will certainly want to pursue the technologies of 'force multiplication.' But they are likely to prove not the means of our salvation from the threat of war, but only means of maintaining the uneasy status quo. With reconnaissance and warning satellites, as with other fields of high-tech military technology, it is likely we shall have to make prodigious advances just to stay in place.

# CHAPTER FIVE

# Force Multipliers: Navigation and Weather Satellites

Operation Eagle Claw, the failed mission to rescue the American hostages in Iran in April 1980, showed how central satellites have become to U.S. military operations. The would-be rescuers in the Joint Task Force carried satellite maps and pictures of Tehran and the Embassy area – materials they had to leave behind in the abandoned helicopters at Desert One. A satellite communications link between Washington, the Task Force headquarters in Egypt, and the deputy commander's C-130 air transport let President Carter pronounce final approval of the decision to abort the mission after the loss of three out of eight helicopters. (By the same link, the Pentagon disapproved the later request of the Task Force deputy commander that *Nimitz* fighter-bombers go in and destroy the remaining helicopters with their 'classified' pictures.)

Earlier, satellite weather pictures (the aircraft carriers have direct readout stations) had indicated no bad weather near the planned helicopter route. But weather satellites can't see *haboobs* – the lingering clouds of fine dust that gave the helicopters a rugged trip, making them an hour and a half late to the rendezvous site and possibly contributing to the equipment failures in two of them.

Even so, if a new satellite system – the Navstar Global Positioning System (GPS) – had been available, one of those helicopters might not have had to turn back, the mission might have gone on to Tehran, and history, for better or worse (depending on what might have happened at the Embassy), might have been different. That one helicopter had lost its gyro and its other radio navigational aids and couldn't safely navigate a 9,000-foot mountain ridge ahead. Equipped with a receiver designed for the new GPS navigational satellite signals, a military pilot would be able to fly through a *haboob* without any other help.

Several weeks before the desert fiasco, the Under Secretary of Defense for Research, William Perry, told a Senate subcommittee about a demonstration he had been given at the Army's Yuma Proving Grounds. 'It was at night,' he said, 'and the demonstration was a blind landing. They brought the helicopter into an airfield using no navigation assistance except the global positioning system which enabled the pilot to locate himself in three dimensions to within about a few feet. The pilot was able to bring the helicopter down. There was a big X on the runway . . . and he landed it literally within a few feet of that X. His only assistance was looking at a needle which was telling him whether to go up or down or left or right.

'Then we went to an operation and we flew out to a normal personnel carrier which was hidden in the brush and flew directly over it, again within about a few feet.' After the chopper put Perry down again, he watched a C-141 transport fly in: '. . . it dropped supplies to a position on the ground. It had nothing to guide it but the global positioning system and even with the errors of wind drift on the parachute drop, those supplies landed about thirty or forty feet from the indicated point on the ground.

'Then, finally, we saw a demonstration of blind bombing using the global positioning system with what we called dumb bombs. Iron bombs with no electronic guidance in them were dropped to within ten to twenty feet of the indicated target with no other assistance again other than the global positioning system.' He pointed out that while the U.S. was developing a 'whole family' of 'smart' weapons with internal electronic guidance systems, each weapon costs tens of thousands of dollars. But, he said, 'There is a technique by simply putting a global positioning system satellite receiver on an airplane where we can convert the ordinary Mark 82, and Mark 84 bombs we have in the inventory to weapons with equivalent accuracy.'

This is the kind of leverage the military call a 'force multiplier.' And it can literally multiply the value of a given set of weapons. For example, one Pentagon-contracted think-tank study concluded that a force of 1,465 fighter-bombers using the GPS could destroy targets at the same rate as a force of 1,714 of the same planes without GPS – and the cost (back in 1980) of the extra 249 planes would be upward of $7 billion. *If* the United States were in fact going to buy those extra planes, then the Navstar GPS would come close to paying for its $8.6 billion cost (in 1980 dollars, through the year 2000).

But there's the rub. If you just measure the costs of the present navigational systems that the Navstar will replace through the year 2000, the savings amount to only about $760 million, less than a tenth of the cost of the GPS. That may be why, in 1981, the House of Representatives refused to appropriate more money to keep the project going. When the House Armed Services Committee voted down the $316.6 million request for the fiscal 1982 budget, the committee chairman Melvin Price said the GPS would be 'nice to have,' but that it wasn't 'essential.'

Thomas Quinn, Assistant Deputy Under Secretary of Defense for C-cubed, said, 'It's much more than "nice to have" – it's an essential program. We consider GPS to be a Number One priority program – it's really up on top of the list. It provides *so* much additional military capability in terms of being able to pinpoint where people are on a continuous basis, and *worldwide*. There is *no* other system that will give you that kind of navigation capability world-wide: if we should have to go into some places in the world today where there is no loran system,* you can go in with portable systems, but nothing that will begin to provide that kind of a service. People I talk to in the Congress say their problem is affordability: it looks as though the Department is initiating programs for a whole lot of things, for which, downstream, bills are going to come in that we haven't anticipated adequately and that we're not going to be able to afford. What we're saying is that, yes, we recognize that that's a potential problem, but we're establishing the priorities and we're going to make sure that

---

* (Lo)ng (ra)nge (n)avigation – uses radio signals from special ground stations.

doesn't happen. But if something has to give, we wouldn't want it to be GPS. And I hope we've convinced them that this is the case and that they will put the money in.'

When representatives from the House and Senate Armed Services committees met to settle their differences on the 1982 Defense authorization bill (the Senate committee had voted for GPS funds, the House committee against), they took Quinn at his word. They instructed the Pentagon to 'reprogram' funds away from other projects to keep the Navstar on its schedule of full operational capability by 1987.

Although there may be room for argument as to whether the GPS is 'essential,' it's hard to deny that it's a technological marvel, an exquisite refinement of one of the earliest military uses of satellites. And the ideas behind Navstar go back even farther, to the first artificial satellite, the Soviet Sputnik, launched in 1957. Observers at the Johns Hopkins Applied Physics Laboratory measured the frequency of the radio signals from the Sputnik as it passed overhead. As the satellite approached, the signal's frequency increased; as it receded, the frequency decreased. This is called the 'Doppler effect,' after the Austrian scientist who first explained it. (Sound does the same thing: as a train approaches, its whistle tone gets higher, as the train moves away, the tone gets lower.) By measuring the Doppler effect and how it varied over a fixed time period, the Johns Hopkins scientists could calculate the speed of the satellite as it passed by, and that calculation in turn, combined with their knowledge of the exact position of their observation point, allowed them to describe the entire orbit of the Sputnik.

Suppose, though, the scientists said, that we *don't* know exactly where we are, but we *do* know exactly what the satellite's orbit is like. Then the calculations could be reversed and the observers could deduce their own position, again by measuring the Doppler shift of the satellite's radio signal. The U.S. Navy was particularly interested in this possibility, because it planned to send Polaris nuclear submarines to sea with ballistic missiles. In order to launch the missiles more or less accurately, the submarines would have to know where they were. In 1958 the Navy hired the

Johns Hopkins lab to design the first Transit navigational satellite.

Building a working navigational satellite system posed a challenging set of technical requirements. First, in order to transmit a regular signal (every two minutes was the period chosen) showing its location, the satellite has to 'know' where it is. A pair of Navy ground stations, one in California, one in Minnesota, 'inject' orbital information into the satellite's electronic memory every twelve hours – or every week in the case of the latest models (the 'injection' stations get that information from four tracking stations, one in Hawaii, three on the U.S. mainland). But that's not good enough: the satellite must also know where it's going to be during the period between data injections, so that every two-minute signal it sends out will be accurate.

Predicting the exact future locations of the satellite would be easy if Kepler's and Newton's laws of planetary motion and gravitation were perfect models of reality – if the earth were a perfect, uniformly dense sphere, if it had no atmosphere, if there were no sun and no moon. But in the real universe, departures from the mathematical ideal complicate orbital calculations. The earth is not perfectly spherical, its mass is ever so slightly lumpy, and the moon exerts an outside gravitational pull. These factors introduce 'higher order' harmonics into the mathematics of orbital motion. Since the beginning of the space age, the Navy, the Army, the Air Force, and NASA have sent several dozen geodetic satellites into orbit. Tracking these and other satellites from a variety of ground stations, combined with years of computer analysis, has allowed the geodesists to build highly refined models of our planet's shape and gravitational field. Those models help make satellite orbits highly predictable. (They have also made it possible for the missile designers to build ever more accurate ICBMs.)

But other natural phenomena as well can disturb the perfect harmony of man-made heavenly spheres. One source of disturbance is air resistance: even at the 600- or 700-mile altitudes of the Transit satellites, there is just enough air left to subtly resist the orbiting object. Even energy from the sun, in the form of 'radiation pressure,' has some effect. These forces too are subject to some mathematical prediction. But the latest generation of Transit-like satellites, called Nova, has equipment for measuring and dealing with the effects of atmospheric drag and radiation pressure. A small ball in a hollow container, inside the satellite at its center of gravity, is shielded from those effects by the outer surfaces of the satellite. The ball, whose position is measured electronically, therefore follows a mathematically ideal orbit. When the satellite's guidance system registers changes in the relative orbits of ball and satellite, it fires gas jets to realign the two.

A satellite also needs some system for 'attitude control' – keeping it faced constantly in the right direction. A variety of methods is available for attitude control, but the Transit series uses one of the more ingenious. The gravity of the earth decreases as a body gets farther from the center of the earth. Over time, this 'gravity gradient' effect can even cause a satellite to start to tumble in orbit, as one end drifts down and sets the whole thing into a slow spin. Now if the satellite is quite long, and if one end is much more massive than the other, the heavy end will stay pointed at the center of the earth as the satellite revolves around it. The Transit series satellite, once in orbit, extends a 90-foot-long boom with a small weight at the end, causing the whole space-craft to stabilize its attitude with the business end pointed to the ground.

Along with mechanisms for a stable and predictable orbit, the navigation satellite requires an especially precise broadcasting system. In particular, if the users of the system are going to measure the Doppler effect correctly, they have to know the exact frequency of the radio signal as it leaves the satellite. This called for unusually accurate crystal oscillators, the devices that determine the frequency of the radio signal. Since the satellite sends a message identifying where it is at the instant of each two-minute time signal, it must also have a highly accurate clock.

That clock keeps time by the same oscillator that governs the broadcast frequencies. The navigation satellite broadcasts in two different frequencies simultaneously, and for a reason. As radio signals pass down through the earth's ionosphere and atmosphere, they are slightly bent and delayed. This effect would throw off the Doppler calculations if it weren't compensated for. But the slowing effect is directly related to the frequency of the radio signal; comparing two signals, then, lets the navigator's computer allow for the non-Doppler changes in frequency.

The Transit family of navigation satellites went 'operational' at the beginning of 1964 and still serves today (it was opened to general civilian use in 1967).* The system does have its limitations. Since only four or five satellites operate at a time, there can be a gap of several hours between the overhead passage of one beacon and the next (the average is an hour and a half). To make use of the Doppler effect, the satellites have to be at relatively low orbits (about 600 to 700 miles), where their speeds will be high. But at lower orbits, enough atmosphere remains to drag the satellite back a little. Moreover, it would take a very large number of satellites to give continuous coverage around the globe. Because the system gives position in only two dimensions, it's useful for ships but not for airplanes.

The accuracy of the system hasn't kept pace with the technology of the latest weapons. One reading from a Transit satellite will tell a submarine captain his position to within 200 to 300 feet; that was good enough when submarine-launched ballistic missiles themselves were only accurate to within a third or a quarter of a mile. But now the Navy wants to shoot for accuracies approaching those of the newest land-based ICBMs – miss-distances approaching 700, perhaps even 300, feet. At such accuracies, the precision with which the launch site is known becomes much more important. Finally, the accuracy of the current satellite navigation system depends on the navigator knowing his exact speed when he takes his

* Geoffrey Perry (of the Kettering School in England and member of The British Interplanetary Society as well as consultant to the Congressional Research Service) and some of his students and colleagues as early as 1976 identified certain Soviet satellites as being navigational. The Soviet series began in 1972, but the Soviets didn't announce their purpose until six years later. The Soviet system, like the Transit, utilizes the Doppler effect, and even uses almost the same radio frequencies (150 and 400 megahertz) as the U.S. system.

measurement: for example, a half-knot error in velocity will lead to a position error not of 200 feet, but 600.*

Even as the Transit system went operational, scientists were studying ways to overcome its limitations. In 1963, The Aerospace Corporation, the Air Force's think tank, began looking for ways to navigate aircraft by satellite. By 1967, the Air Force had let contracts to Hughes Aircraft and TRW Systems to design such a system. The Aerospace group concluded that a complex of twenty geosynchronous satellites would be best. Instead of utilizing the Doppler effect as Transit did, the new system would use time-ranging. The speed of light (and other electromagnetic waves) is a constant 186,300 miles per second: if you can measure the amount of time a radio signal takes to get from a satellite to you, you can calculate how far away from it you are.

Meanwhile, the Naval Research Laboratory had been working on building super-accurate clocks into navigation satellites – the kind of clocks that would be needed for time-ranging using the speed of light. In 1967 the Navy sent up the first Timation (from Time Navigation) test satellite. In 1973 the Air Force and the Navy worked out a compromise between their designs, merging their separate programs into the Global Positioning System (also some-

---

* Strictly speaking, missile accuracies are estimated in CEP – circular error probable. CEP is the radius of a circle within which half the missile warheads will land. Missile inaccuracy is determined by a combination of factors called the 'error budget.' Over the years, research has brought these factors (which include the variations in the earth's gravitational field, the technical limitations of the missile's guidance system, the imprecision of rocket fuel burning rates, the effects of the atmosphere on re-entering warheads, and the exactness of estimates of the relative locations of launch site and target) under greater control of the missile designers. Therefore, relative to the total error budget, the contribution to the missile's error by mis-estimates in the firing location of the missile has become much greater. The more accurately the missile guidance system knows where its launch point is, the closer it can steer the missile to its target thousands of miles away.

Another new satellite, the GEOSAT, will help the Navy further improve missile accuracy. The GEOSAT, to be launched late in 1983, will carry a radar altimeter to measure the distance between the ocean surface and the satellite. This information will allow Navy scientists to make better calculations about how far the earth varies from being a perfect sphere. Those calculations, in turn, will let them build more accurate models of the local gravitational variations in those regions of the ocean from which the Navy would launch ballistic missiles.

times known as Navstar – for Navigation System using Timing and Ranging). The Air Force (through what is now known as the Space Division) took over the overall management of the program; the Navy continued research on the satellite atomic clocks with a series of Navigation Technology Satellites. Rockwell got the contract to build the Global Positioning System satellites.

The heart of the Navstar satellite is an atomic clock (three of them, actually, for redundancy). By referring to a rubidium-atom oscillator that is stable to within one second in thirty thousand years, the satellite clock remains accurate to a millionth of a second. (The Navy is experimenting with a 'hydrogen maser' oscillator to put on the satellites: it could be stable to within one second in *three million* years.) By comparison, if you have a quartz-crystal electronic watch, it is probably accurate, at best, to about one second in a few days.

All the satellites (there are now about half a dozen; the system is supposed to be fully operational with eighteen by the end of 1987) will be synchronized into a common 'GPS time.' Each satellite will broadcast a time signal and its exact location at the time the signal goes out. By measuring the difference between the time the signal was sent and the time received, then multiplying by the speed of light, the navigator's receiver-computer gets the apparent distance between himself and the satellite. If the receiver tuning in on the GPS had its own atomic clock on GPS time, it would need the distances to three satellites at one time to triangulate its position in three dimensions. But it isn't feasible for each receiver to have its own costly atomic clock; instead it will take time and range signals from four satellites at once. This allows the receiving computer to compensate for the bias, or error, in its own, less precise clock. The computer will be able – as the mathematicians put it – to solve 'four equations with four unknowns' (the four unknowns being the three dimensional coordinates of the receiver's location plus the clock bias). Although the receiver will use 'time-ranging' to find its location, it will also be able to make use of the Doppler effect: in this case, not to determine position, as with the transit system, but rather to calculate velocity – within a tenth of a mile per hour – in addition to position.

The time-ranging principle on which the GPS operates is fairly straightforward. Engineering that principle into a working military system is another, more intricate problem. The biggest job is shaving off as much of the potential error in determining satellite positions as possible. Like the earlier Transit satellites, the GPS satellites transmit on two radio frequencies simultaneously, so that the receiver can calculate the effects of the ionosphere in delaying the signal. Einstein identified another source of error in his general theory of relativity: gravity slows down time itself. So precise are the atomic clocks that the difference in gravity between the satellite 12,500 miles up and the receivers on the earth's surface, that the satellite's oscillator, or timekeeper, is slightly offset to compensate for the difference in satellite time and ground time.

A sophisticated ground control system will keep up to date the time and location messages the satellites send out. A Master Control Station, now at Vandenberg Air Force Base (probably to be moved to the new Consolidated Space Operations Center in Colorado Springs), will run the operation. As the system now works, a monitoring station at Vandenberg and three others – on Guam, in Hawaii, and in Alaska – constantly read the signals from the GPS satellites passing overhead and use their own atomic clocks to calculate the satellite positions. These remotely controlled stations also record information on the local weather conditions so that the Master Control Station can calculate the effects of atmospheric delay on the satellite signals.

The Master Control Station sends the collected information to the Naval Surface Weapons Center in Dahlgren, Virginia. The Navy computers predict the future locations – the exact orbits of the satellites. Then the Master Control Station, using these predictions as guidelines, makes last-minute estimates, based on such factors as the probable immediate changes in pressure on the satellites from the sun's radiation, of each of the satellites' positions in the next twenty-four hours. At this point the Upload Station, also at Vandenberg, feeds the predicted locations into each satellite's memory as it passes over. Meanwhile, the Air Force Satellite Control Facility, through its headquarters in Sunnyvale and its worldwide Remote Tracking Stations, is also watching over the satellites, performing the 'station keeping' and 'housekeeping' tasks that keep the satellites in precise orbits, properly oriented, and in working condition (the satellites are designed to work for at least five years, and they carry enough fuel to last seven).

Once the system is fully up and operating, the 'ground segment' will be configured this way: the Master Control Station will be in Colorado; four other Monitoring Stations will be on Guam, Diego Garcia (in the Indian Ocean), Ascension Island (off Africa in the Atlantic), and Hawaii; and three 'Ground Antennae' – in the Philippines, on Diego Garcia, and on Ascension Island – will 'upload' the latest data to the satellites.

The purpose of all this effort, of course, is to give U.S. military forces a useful navigation system. The Air Force intends to buy almost 12,000 GPS receivers and install them on just about all its aircraft. The Army will buy at least 1,600 receivers, but perhaps as many as 13,000, depending on other position-finding systems it has been considering. The Navy will buy at least 1,400 sets, but possibly as many as 4,700. Magnavox has had the main contract for developing these user sets.

The GPS has the virtue that it puts all these military users in a 'common grid' of time and space – they can all place themselves on the same three-dimensional map encompassing the seas, the land, the air, and the space around the earth. The list of possible uses of this revolutionary development in navigation grows constantly. The tests that Under Secretary of Defense Perry reported – helicopter blind landing, helicopter rendezvous with a land vehicle, blind airdrop of cargo, ultra-accurate bombing – reflect only a sampling of the possibilities.

Because there is only a handful of satellites in orbit so far, most of the GPS tests have just been demonstrations. But the Navy has already made a very practical application of the system. It's using the Navstar in Trident submarine-launched missile tests to increase the accuracy of the missile. According to Rear Admiral Robert Wertheim, then Director of the Navy Strategic Systems Projects Office, the Navy researchers have set up a system they call Satrack: the Navstar signals '. . . transpond through a device in the missile which then retransmits to ship and shore station signals from which we can precisely reconstruct the missile's position and its velocity in flight on an instant-by-instant basis.

'With that kind of full trajectory tracking data,' said Wertheim, 'we can conduct flight tests from any place in the ocean, from the Atlantic or the Pacific, and get trajectory data sufficiently accurate to be able to identify the specific source of errors; for example, the accelerometers, the gyros, or initial condition errors for the missiles.' The ultimate purpose of these tests is to make sea-based missiles as accurate as the land-based ones – so they too can attack hard-to-destroy Soviet missile silos in 'counterforce' scenarios of nuclear war. The GPS will contribute more directly to that goal, of course, by making it possible for the missile-launching submarines to keep track of their exact positions at sea.*

On a more mundane level, the GPS should improve day-to-day naval operations. Ships will be able to navigate safely in and out of fogbound harbors. In a man-overboard rescue, a ship could return to the place where the man was lost – or precisely retrace its recent course. Airborne search and rescue would also be easier, the searchers scanning a 'common grid' with high precision. Navstar should also make Navy anti-submarine warfare operations easier: air and surface submarine hunters will be able to coordinate their searches more closely, and one could

* Why not put Navstar receivers on the ICBMs and sub-launched missiles themselves, using the satellites as part of the missile guidance system? The Air Force in fact considered this possibility for its new MX intercontinental missile, but in the end decided against it. With its new advanced inertial (and therefore self-contained) guidance system, the MX will be almost as accurate as it would with GPS, and in any case accurate enough to hit almost any Soviet target with a 99 percent probability of kill. In the early stages of a nuclear war, the Air Force would prefer not to take even a remote chance that Soviet jammers could interfere with the missile's GPS reception; in the later stages, the ground system supporting the GPS would probably be gone and the accuracy of the satellite beacons would begin to decline.

expect the Navy's multi-ocean listening sensors to be more accurate when they share the GPS common grid. Navy carriers and their aircraft should find their operations helped as well. And when Navy or Air Force aircraft want to rendezvous at long distance, the GPS will make it a snap. That includes the rendezvous of fighters with aerial refueling tankers. Since the GPS is a 'passive' system – the receivers tune in on the satellites, but don't have to emit any radio signals themselves, or don't have to communicate with each other to rendezvous – military ships and planes units will be able to operate in radio silence and avoid giving themselves away to enemy electronic warfare units.

Land forces should also find the Navstar useful. 'What a tremendous asset,' said General Henry, head of the Air Force Space Division, '. . . to make a landing on a beach and not have to put up flags or send people down ahead of time to tell you where to come in.' Helicopters evacuating wounded troops could do so more quickly and safely. Ground forces calling in air strikes should be able to give exact target coordinates – and have less worry of their own planes accidentally hitting them. Long-range artillery should be more accurate with GPS. Commando operations behind enemy lines would have the same benefit.

Military ingenuity will no doubt find still more uses for Navstar. For example, the 'time transfer' feature of the system – the ability it gives users to synchronize their clocks with incredible precision – will be valuable to electronic warfare specialists. The principle they will use will be something like the GPS time-ranging one: if you can figure out how long a radio signal took to arrive from the source, you can figure the distance to the source by multiplying the time by the speed of light.

Some electronic warfare listening posts are on ships and planes, but many are fixed stations on the ground. These stations tune in to enemy radio or radar signals, perhaps to get information from them, perhaps to prepare to jam them, perhaps to plan an attack on their source. If four or more such stations can each intercept the same signal at the same time, and if they can measure the exact time at which

The Boeing Inertial Upper Stage pushed the state-of-the-art for solid-fueled, self-navigating rocket stages. The Air Force and NASA originally envisaged the IUS as an "Interim" Upper Stage, to be used until a full-fledged "Orbital Transfer Vehicle," or space-tug, came into operation.

The Orbital Transfer Vehicle (OTV) would have been a manned space ship carrying heavy cargoes from low earth orbit to the geosynchronous orbit and back. At space stations in the lower orbit, pieces of large platforms or space stations for the higher orbit would be assembled, then ferried up. Above is a cut-away view of a hangar at a low-orbit space station sheltering a Boeing-designed "Future Orbital Transfer Vehicle" between flights. BOEING AEROSPACE COMPANY

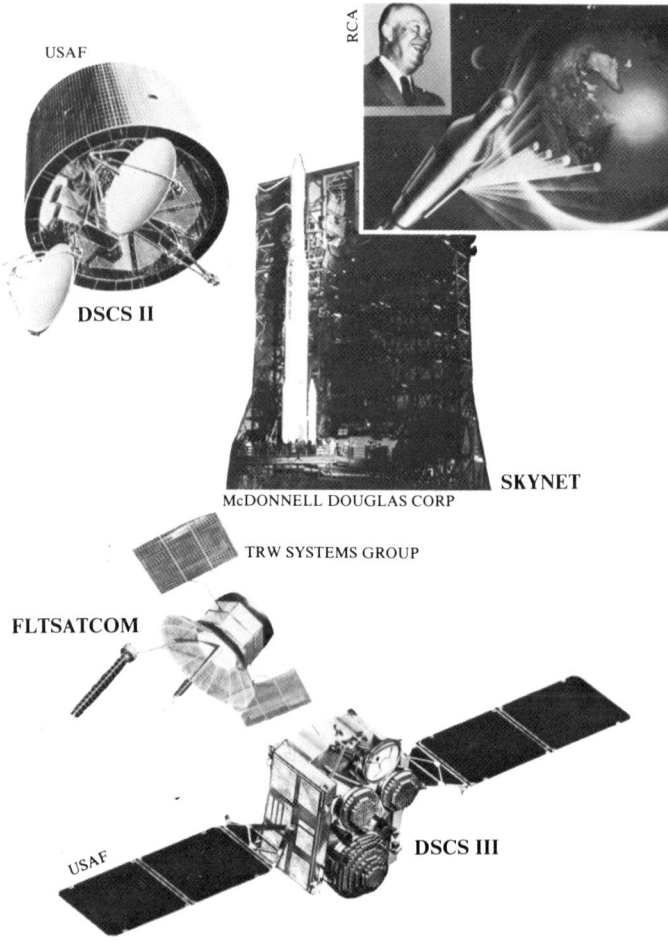

USAF

DSCS II

RCA

McDONNELL DOUGLAS CORP

SKYNET

TRW SYSTEMS GROUP

FLTSATCOM

USAF

DSCS III

The first geosynchronous military communications satellite was the British Skynet. Far more capable was the later U.S. Defense Communications Satellite Phase II, or DSCS II. The four operational DSCS II satellites each relay a hundred million bits of digital data per second, carry 1,300 two-way voice circuits at once over Super High Frequency radio channels. The DSCS II satellites have been the backbone of the World Wide Military Command and Control System. The Navy's Fleet Satellite Communications System Satellites (FLTSATCOM) can send messages to about 600 ships around the world.

The newest addition to the U.S. military communications satellite series is the General Electric-built DSCS III, the first of which went up on the fall of 1982. It is designed to resist the radiation effects of nuclear weapons and to defeat Soviet radio-jamming efforts.

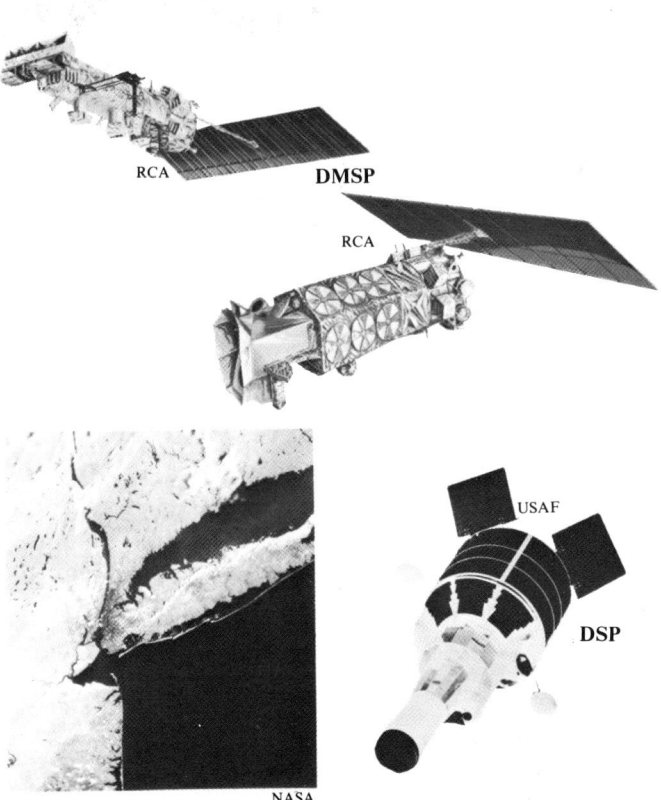

RCA **DMSP**

RCA

USAF

**DSP**

NASA

The Block 5D-2 Defense Meteorological Satellite Program spacecraft is the latest in the Air Force series of weather satellites. The DMSP weather sensors ride a platform similar to that of the National Oceanic and Atmospheric Administration's Tiros-N series, and both types of satellite travel in a "sun-synchronous" orbit, passing over the equator at the same local time of day on each revolution.

The KH-11 imaging reconnaissance satellite probably carries multispectral scanners similar in principle to the one on Landsat 4 which took this picture of the New York City area from 115 miles up. But where the Landsat scanner records images with a ground resolution of about 260 feet, the ground resolution of the KH-11 images is probably about 20 to 30 times better.

The Defense Support Program early-warning satellite scans the earth with a twelve-foot long infrared telescope, sensitive to the exhaust plumes of rocket launches. The three DSP satellites keep the North American Aerospace Defense Command immediately informed of any rocket firings, and in the event of war could render an approximate count of the number of Soviet missiles on the attack. The satellite also carries sensors to indicate the occurrence of nuclear explosions.

The Air Force's Maui Optical Tracking Station photographed the fiery trail of Soviet ocean reconnaissance satellite, Cosmos 954, as it plunged into the atmosphere on January 28, 1978. This type of Soviet spacecraft uses a nuclear-powered radar to scan the oceans for U.S. ships. USAF

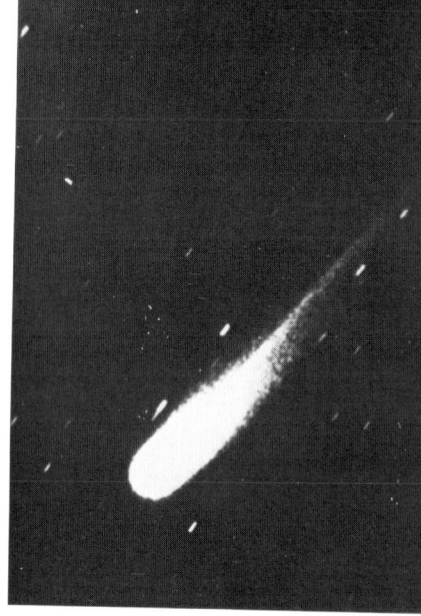

The Air Force is moving toward more "real-time" surveillance of the earth from space. The "Teal Ruby" experiment, launched from the space shuttle in 1983, was designed to lead the way toward a satellite system that uses infrared sensors to detect and track aircraft flying 460 miles below. Interceptors can be dispatched toward an enemy bomber almost as soon as it takes off. USAF

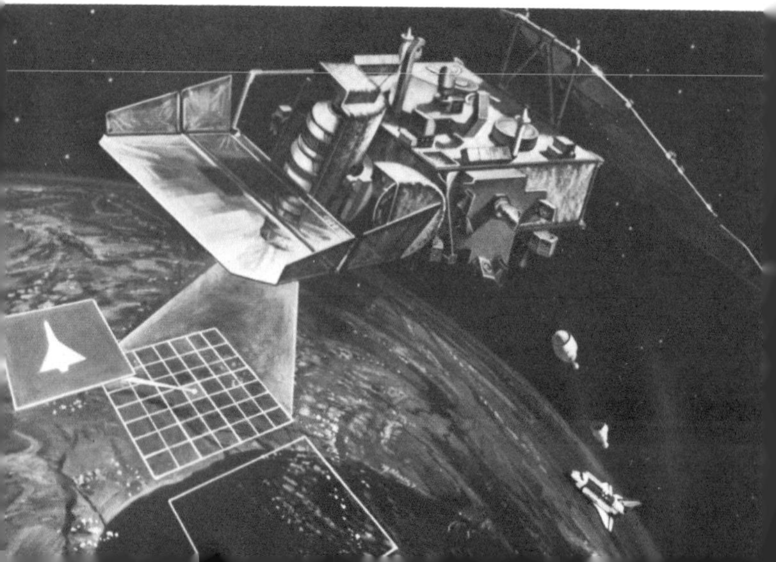

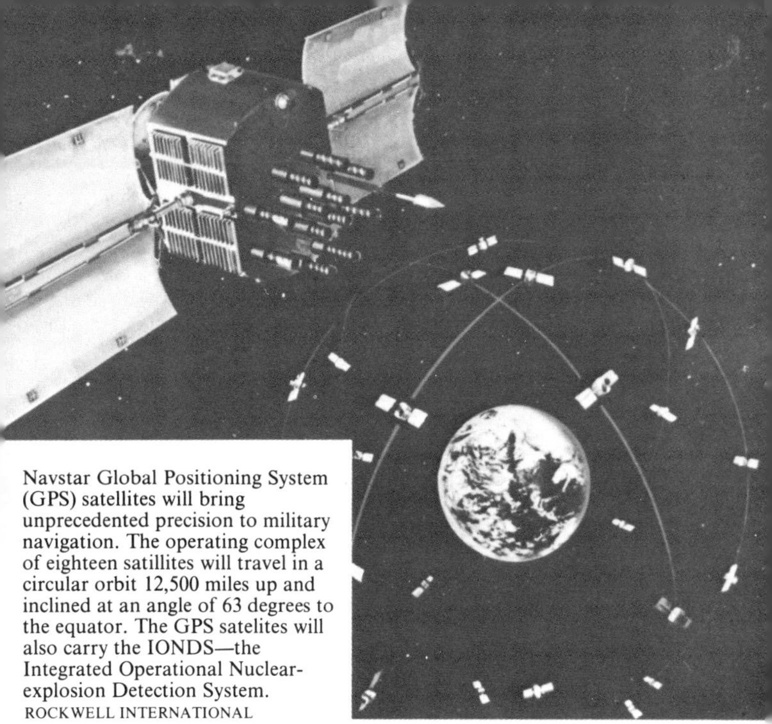

Navstar Global Positioning System (GPS) satellites will bring unprecedented precision to military navigation. The operating complex of eighteen satillites will travel in a circular orbit 12,500 miles up and inclined at an angle of 63 degrees to the equator. The GPS satelites will also carry the IONDS—the Integrated Operational Nuclear-explosion Detection System. ROCKWELL INTERNATIONAL

Two soldiers are seen using the GPS manpack navigation units in a field exercise. The manpacks weigh 27 pounds and can determine position within 45–50 feet and velocity within fractions of a mile an hour. USAF

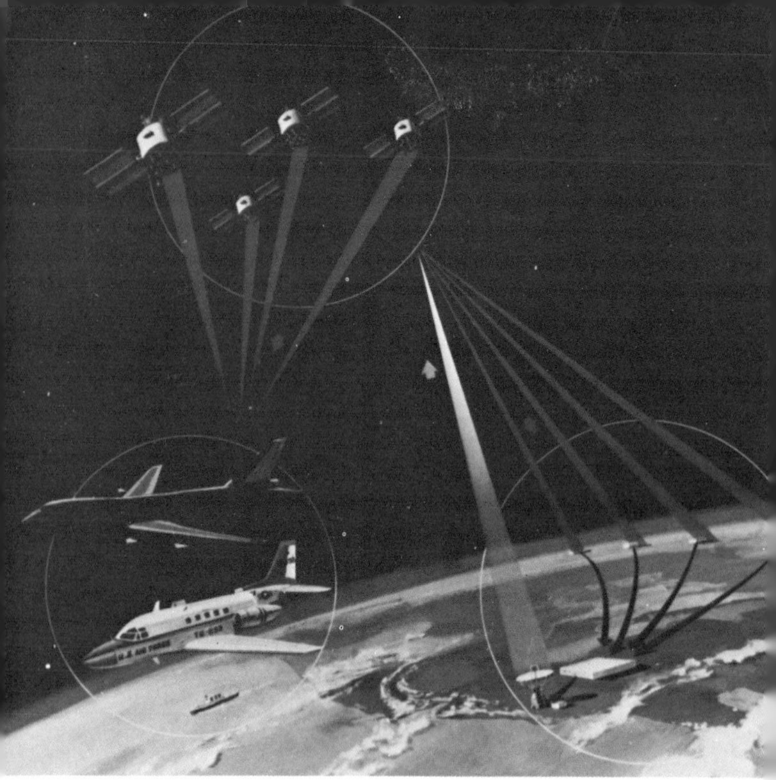

In the Global Positioning System, the ground stations of the Control Segment will monitor the satellites and feed them information locating them precisely in space and synchronizing their on-board atomic clocks (accurate to one second in 30,000 years). The satellites in the Space Segment of the system will send down signals indicating their distances from the using receiver. Computing the position and timing data from four satellites at one time, the GPS receiver unit will indicate its own exact location and velocity. ROCKWELL INTERNATIONAL

The late satellites of the GPS will be carried into space by the space shuttle. The GPS will be useful not only to terrestrial vehicles but also to spacecraft as well. ROCKWELL INTERNATIONAL

The Miniature Homing Intercept Vehicle (MHIV) is the heart of the new U.S. anti-satellite weapon, being tested in space first in 1983. The F-15-launched system is called the Prototype Miniature Air-Launched System, or PMALS.

The two-stage rocket which will carry the Miniature Homing Intercept Vehicle of the PMALS anti-satellite is mounted under the body of an F-15 fighter by Boeing technicians in a test "integration" of the system. USAF

The 12″ by 13″ cylinder spots the target satellite through eight infrared telescopes clustered at the center. A laser-gyroscope and a tiny computer determine what final course adjustments are necessary. Around the outside of the MHIV are 56 small rocket tubes which fire out nozzles belting its middle. Precise firings of these rockets make sure the vehicle stays on a collision course with the target. VOUGHT CORP.

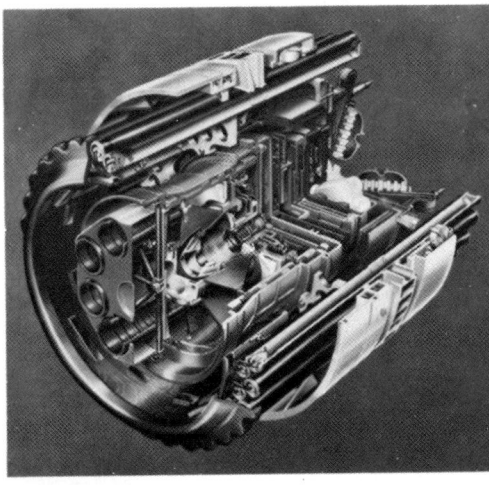

The MHIV gets a 7,500-mile-an-hour boost from a two-stage rocket, which in turn has been carried to high altitude by ab F-15 fighter plane. The Space Command's Space Defense Operations Center in Colorado will identify the targets and send firing instructions to the F-15 pilots. USAF

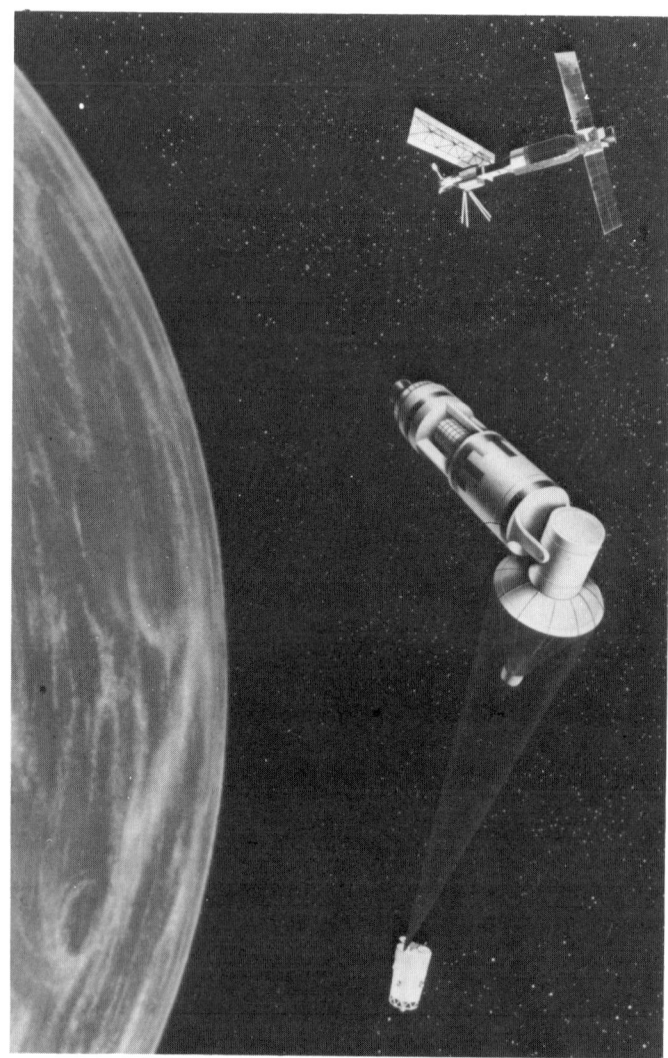

Some believe that the weapon of the future is the space-based laser battle station. In this artist's conception, a large mirror directs a laser beam against an enemy satellite, perhaps defending the friendly satellite at left. More ambitious schemes envisage complexes of thirty or more even larger laser battle stations able to attack Soviet ICBM's and submarine-launched missles as they rise into space. DEPARTMENT OF DEFENCE the signal arrived, they can compare times of arrival. From that information, the electronic intelligence analysts can triangulate in on the signal's 'emitter.' But their ability to locate the source that way hinges on the perfect synchronization of the clocks in the stations measuring the times of arrival of the common radio signal. If the stations can coordinate their common time base to within a hundred nanoseconds – a hundred *billionths* of a second – they can track the signal back to within several hundred feet of its source. (Light travels at about 983 feet every millionth of a second.) Navstar will let them do that by locking into the GPS time grid.

The Global Positioning System should prove useful not just on land, in the water, and in the air, but even in space. Satellites might do their own 'station keeping' by adjusting their orbits in reference to the GPS grid. Rockwell and NASA will be outfitting future space shuttle orbiters with GPS receivers. The shuttle could then enter orbit, change orbits, rendezvous with other satellites, and land without depending on foreign-based ground stations. Precise position location in space is going to be particularly important when the Air Force begins to use the shuttle as a reconnaissance platform for imaging objects on the ground or in space. Of course the GPS could improve the precision of unmanned spy satellites too. (The latest U.S. civilian land observation satellite, Landsat 4, carries a GPS receiver.)

Besides performing its navigation and timing functions, the GPS is going to provide another, less well-advertised service. The Navstar satellites will carry nuclear explosion detectors something like those on the Vela Hotel and the early warning satellites. These detectors will be linked into a system called IONDS – the Integrated Operational Nuclear Detection System. The official Air Force Space Division and Systems Command handouts on the GPS don't even mention IONDS. Semiofficial articles in military industrial journals make general references to the usefulness of IONDS in detecting any violations of the 1963 treaty banning nuclear weapons tests in the atmosphere. But the real enthusiasts for IONDS are the nuclear war command-and-control planners.

What the strategic nuclear war planners are interested in is 'damage assessment.' In a nuclear war, it would be of some value to be able to figure out quickly how many Soviet nuclear weapons went off and where in the United States (or in Europe, if the war were somehow limited to that theater). But it would be even more useful to get instant reports on where *U.S.* nuclear warheads had successfully detonated in the Soviet Union. Then the U.S. commanders wouldn't waste any of their follow-on weapons on targets that had already been destroyed. Since each weapon is less likely to be wasted, each in a sense becomes worth more – fewer weapons are needed to be sure of accomplishing the same amount of destruction.

This possibility becomes even more intriguing when you consider the option of attacking the other side's nuclear missiles in their silos, before he has a chance to launch them. The missiles and their warheads that the attacker might send in the first wave would not be 100 percent reliable – some would inevitably fail to reach their targets. While it's possible to estimate the approximate total number of unreliable weapons, it's not possible to predict in advance *which ones* will actually fail. The only way now to hedge against the failures is to aim at least two warheads at every target, hoping that at least one will make it through. But with immediate feedback through IONDS, the strategist's computers could figure out where warheads did go off, and therefore where they *didn't* go off. Then the missilemen need only aim a single warhead at each target in the first wave, following them quickly with a second wave aimed only at the missed targets. With IONDS, a 'first strike' intended to disarm the other side's ICBMs looks feasible with a far smaller arsenal than before.

So Navstar also needs to be looked at in the context of that trend we saw in the communications field: the growing interest of military planners in preparing to fight 'limited' or 'escalation controlled' or 'protracted' nuclear wars as though they were just more challenging versions of the conventional wars that history has familiarized us with. Will they start to believe they can pull it off?

Militarily, the GPS seems to have something for everybody. But the system also has a wide range of potential civilian applications. Civilian as well as military users of the shuttle could call on the system. It could improve maritime navigation both on the high seas and in harbors and narrow sea-lanes. Rockwell suggests that the ability to find exact locations repeatedly should help the oil companies search for oil and gas. The company also argues for the usefulness of the system in geologic research, geodetic surveys, and aerial mapping projects. And, in theory, at least, the GPS could revolutionize the whole air traffic control system.

But when it comes to civilian applications, there's a hitch. The military obviously doesn't want to let potential enemies get the same positioning accuracy out of the GPS as the United States will. So the publicly available Navstar channel (the other will be in virtually unbreakable code) will only permit accuracies to about 656 feet – and that only at the '50 percent confidence level.' Under Secretary of Defense William Perry said in 1980, 'Some people argue that even at the degraded level, we're giving them [the adversary] a very significant advantage, having that world-wide capability twenty-four hours a day. This is a far better capability than they can get with any navigational systems that exist today.' The 'clear' or 'coarse acquisition' accuracy went down to that level when the Pentagon decided to save money by going from a constellation of twenty-four satellites to the currently planned eighteen.

The FAA complained that the 656 feet wouldn't even be as good as the present land-based radio navigation system. Depending on just how the satellites are finally placed relative to each other, there is still a chance that the accuracy available at least over the United States could approach what the FAA wants. Meanwhile, the FAA and civil aircraft users are not quite ready to spend the billions that would be necessary to convert their electronic equipment from the current systems to GPS.

With at least a lot of potential in the civil economy, and with a lot of promise as a military 'force multiplier,' why hasn't the GPS evoked more support – and money – from Congress? The Navstar may be a perfect example of the spacemen's lament: like other military space programs, it lacks an organizational constituency, an 'advocacy base.' It does a little something for everybody, but not enough for anybody. The president of The Aerospace Corporation, Eberhardt Rechtin, said, 'One should be able to predict a fine future for such a system. However, it has a problem common to all systems that broadcast to large numbers of passive users: who pays for the transmitter (satellites)?'

When Congress voted on Navstar, the Strategic Air Command was to have control of the system (now likely to go to the new Space Command). But when it comes to budget fights, SAC is more concerned with things like the B-1 bomber and the MX missile. The Air Force Space Division, in charge of developing the GPS, has other pressing demands as well (not to mention its secondary role in the final operation and use of the satellites). And none of the program managers whose budgets had to 'eat' the reprogramming of funds for GPS that Congress ordered for 1982 could have been very happy about it. The generals of the Tactical Air Command might be impressed by the idea that the GPS would let them do more with fewer planes, but they just might like to have the extra planes instead.

Then there's the Not Invented Here syndrome. Example: The Army R-and-D chief, Lieutenant General Donald Keith, agreed in 1981 that 'the pace of battle in future conflict will demand that combat commanders be able to precisely locate themselves and other friendly units.' But he wasn't making a pitch for the GPS; instead, he was touting the Army's own (with the Marines) Precision Location Reporting System (PLRS), which would allow 'combat units to navigate the battlefield in all weather conditions and automatically determine the location of other friendly units.' The PLRS will be ground-based and, unlike the GPS, automatically report to the commander where his troops are. The Army estimated in 1980 that if it didn't get the PLRS, it would need more than 13,000 Navstar receiver units. But with the PLRS, it wouldn't need many more than 1,600.

A further difficulty for the GPS in getting advocates in the Pentagon and supporters on Capitol Hill has been the problem of quantifying its benefits. One firm legacy of the controversial McNamara reign at the Pentagon has been cost-effectiveness analysis. An example of that kind of analysis was the study showing that a given number of fighter-bombers equipped with GPS could do the same job as a larger number without GPS. But in general, it has been difficult to quantify the advantages of the GPS over its alternative: continuing the current navigational systems. The advantage has to come not in the cost of operating the systems themselves (the current systems win hands down in that comparison) but in the 'force multiplication' benefits that the new GPS capabilities are supposed to provide.

It seems obvious that worldwide, continuous position location will save money and increase the effectiveness of the vehicles using the system. But how much money? By what factor will the forces be multiplied? Since the air, ground, and sea forces haven't had a global common grid before, who can say just how much more powerful the GPS will make them? The qualitative improvements may be hard to quantify, but the budget-makers want numbers. And the Congressmen want to know the bottom line in dollars.

No discussion of force-multiplying 'support' systems in space would be complete without mention of weather satellites. In fact, weather satellites have a lot in common with navigation satellites. Both got early attention from military space researchers. Both improve military performance by providing information on the environment. Both went operational in 1964. The new Space Command runs the Defense Meteorological Satellite Program (DMSP) as it will the GPS. And each at times has been the subject of Congressional skepticism.

What the Congressmen wanted to know was if they were going to finance a sizable civilian weather satellite program, why would they have to do it all over again for the military? The way the Air Force handled that one in the early sixties was to low-profile it. As Bernard Schriever, who was then in charge of Air Force space programs, said, '. . . we kind of had to bootleg the defense weather satellites.' That is, the Pentagon didn't admit, except to those with a 'need to know' classified information, that it had its own weather satellites. If you go back and look at the public satellite logs, and you know what to look for, the military weather satellites stand out – a couple of launches a year in the latter half of the 1960s. How the military managed to 'bootleg' its weather satellite program in the sixties is fairly obvious: it was necessary to the satellite photo reconnaissance program. You could waste a lot of film taking pictures of clouds. The weather satellites reconnoitered for the spy satellites.

Two kinds of orbits are most useful for weather satellites. The pictures you see on the evening news come from satellites in the geosynchronous orbit at 22,250 miles, where the cameras stare down at their assigned patches of the globe. Those satellites show the large-scale trends, but don't give the detailed information the meteorologists need. So both the National Oceanic and Atmospheric Administration (NOAA) and the Air Force operate weather satellites in a separate type of orbit, the 'sun-synchronous.' Passing over the poles, at altitudes around 500 miles, these satellites 'see' different strips of the earth at the same local time every day. The orbit traces a line on the ground that moves around the earth at the same rate as the line between day and night. In other words, the angle between the plane of the orbit and the line from the earth to the sun is always the same. If the satellite crosses the equator at sunrise on its first trip north, then from the point of view of the satellite, it is always sunrise on every northward pass over the equator. The NOAA and Air Force satellite orbits have periods of about 101 minutes, so they make complete circuits of the earth 14.3 times every twenty-four hours. The combination of this orbit and the earth's rotation works out so that each satellite passes over the whole surface of the planet twice a day. By the same token, any earth station taking data from the satellite will find it in line of sight twice a day.

Air Force Colonel Stephen McElroy, who works on space shuttle planning at the Pentagon, and previously worked on the meteorological satellite program, knows the differences between the civilian and military versions. 'There is a free flow of satellite data between Global

Weather Central and NOAA,' he said. In fact, Air Force Global Weather Central at Strategic Air Command headquarters outside Omaha and the NOAA Satellite Operations Control Center at Suitland, Maryland, both have satellite communications terminals that link them together via RCA Satcom.

McElroy points out that despite the regular interchange of information between the two weather satellite systems, 'The missions are significantly different. The DoD has, let's say, strike-plan missions, so we need to know what the weather is right at dawn. So we fly a satellite right along the night-morning terminator. It crosses the equator 14.3 times a day, and that data is transmitted in real time to users spread throughout the world – the ones that are in line of sight to the satellite as it passes overhead. The local commander makes his decision on launching his airplanes, or where to drive his carrier to recover, or whatever. And then we have a satellite that flies along about noon that not only helps the folks that got the early morning satellite data confirm their forecasts, but also prepares for the afternoon thunderstorm buildup and such things so you can make your predictions for the rest of the day. We like to have the satellite crossing the night-day terminator as it goes over the North Pole.'

The Global Weather Central reads out the taped DMSP satellite data from its two main ground stations in Maine and the state of Washington. But the satellites are also equipped with APT – Automatic Picture Transmission – so that other terminals can get pictures directly, without going through Global Weather Central. Several U.S. Navy aircraft carriers have terminals for that purpose. The Harris Corporation builds the latest type of mobile military weather satellite terminal, the Mark IV. The Mark IV consists of a van, a power generator, and a 10-foot dish antenna. The generator and the antenna can be packed up in the van, and the whole kit loads onto a C-130 military transport plane. The Mark IV will pick up pictures both from the DMSP and from the NOAA satellites, electronically add grids and labels to them, and let the meteorologists pick out and enlarge any piece of the picture they choose.

Like the other space 'support' systems, the DMSP weather satellites play a role in U.S. readiness to fight a nuclear war. Global Weather Central is under the Strategic Air Command. SAC wants to know at all times what weather its B-52 and FB-111 bomber pilots would run into if they had to make their runs into the Soviet Union. And the Command also keeps its Minuteman ICBM guidance systems so precisely programmed that they can take into account the weather conditions their nuclear warheads would encounter as they re-entered the atmosphere over the Soviet Union from space. In 1981 the Strategic Air Command requested, but didn't get, money for a secret study on how the DMSP could be made more useful in a nuclear war.

McElroy explained how the NOAA and DMSP satellites differed.' 'It turns out that the NOAA folks have a completely different set of requirements. We want visual data. Even our models of global weather are built around visual images – infrared images of clouds: we built our whole system on that for military applications. For the NOAA folks, the images are ancillary: they're in the sounding business. All their models are built on soundings. They're concerned about frontal systems, and the sort of things that you do to predict weather thirty-six hours in advance, and to provide data to all sorts of users around the world. We're a little more tailored. Admittedly, they fly an imager, but their imager is not near[ly] as good as ours – and doesn't cost near[ly] as much. They don't need the additional resolution, so they don't want to get involved in buying our imager. If, on their satellites, their imager dies, they don't launch another satellite, because their primary mission is sounding.'

In the mid-1970s NOAA and the Air Force decided they could use the same basic spacecraft 'bus,' built by RCA. The NOAA version took on the name Tiros-N, while the Air Force version was called Block 5D. The 'bus' is a platform for keeping the satellite in precise orbit, keeping its sensors aimed in the right direction (that is, controlling the 'attitude' of the satellite), and keeping the equipment at the right temperature and supplied with adequate solar and battery power. The latest Air Force version, the Block 5D-2, is 14 feet long, 5 feet in diameter, and weighs about 1,500 pounds (400 of that being the actual sensors).

Like the KH-11 spy satellite scanners, the Block 5D imager (built by Westinghouse Electric) has a telescope that sweeps from side to side, at right angles to the satellite's orbit. The swath covered by the imager is about 1,800 miles wide. The light the telescope gathers registers on electronic detectors in the telescope focal plane, and these record the light signals as bits of digital information. The satellite can store the information on tape or pass it on directly to terminals on the ground or at sea. When the ground terminal's computer pieces the digital bits back together, they form images of the area the satellite has passed over. The daylight images have a resolution of about 1,800 feet (i.e., they show individual clouds as small as 1,800 feet across); moonlight images have a resolution of less than one and three-quarters of a mile. Images in infrared light have a resolution of 1,800 feet, day or night. As McElroy pointed out, the NOAA imagers produce pictures with less resolution: two and a half miles for some ground stations, six tenths of a mile for others, depending on how good their equipment is.

Though NOAA may be more concerned with atmospheric soundings, the Block 5D satellites also carry 'sounding' and other sensors: a scanning infrared radiometer that measures vertical differences in temperature and water vapor in the air, a passive microwave temperature sounder that 'sees' through clouds to measure temperature from the ground up to more than 18 miles' altitude, an ultraviolet sensor that measures the density of the atmosphere between 60 and 150 miles (important in operating spy satellites), an electron spectrometer that helps predict the patterns of the aurora borealis – the northern lights – which can affect radar and radio transmissions, sensors to measure ions and gamma rays, and an ionospheric sounder (important because high-frequency radio transmissions reach long distances by bouncing off the ionosphere).

The next major change in the DMSP satellites will be an even more sophisticated sounder: it will actually be a microwave imager – a sensor that can form pictures from the microwave energy passing up through cloud cover. The same instrument will gather data about rain, wind speed, soil moisture, and sea ice. Hughes Aircraft is building the sensor and the Air Force will launch it from Vandenberg in 1984.

Colonel McElroy argued that it would be difficult to combine the NOAA and Air Force programs. It may look as though both types of satellites are flying the same orbit '. . . but the ascending node times are nowhere near the same.' (One of the NOAA satellites crosses the equator going *south* at 7:30 in the morning and the other crosses going *north* at 3:00 in the afternoon.) Their models of the atmosphere are based on those different times of data collection. 'Now can we combine and reduce the costs? Probably. Every two years is the gestation period: we give birth to a "Lets-go-and-decide-if-NOAA-and-DMSP-can-combine-missions" study. But it would take three satellites to do it – at a minimum – maybe even four. And it's not clear that there would be any large benefit to combining them, except that NOAA's under terrible budget pressure.'

The Soviet Union has avoided this whole problem: it seems to have essentially one meteorological satellite program, and although it's nominally 'civilian,' clearly the Soviet military makes regular use of it. The Soviet Meteor series now seems to consist of two types of satellites. One of these has an orbital inclination of 81.2 degrees and an altitude around 530 miles; the other has an inclination of 97.9 degrees and an altitude around 370 miles. These satellites have sensors pretty much similar to those on the American versions, as well as Automatic Picture Transmission, so that terminals anywhere in the world can read out pictures directly from the satellite as it passes overhead.

So far we've looked mainly at the 'support' systems, the 'force multipliers,' the space systems that help the land, sea, and air forces do their jobs better. It's hard to think of a kind of military operation that doesn't depend on getting information of some kind from satellites to some extent. And military dependence on satellites is growing. But the more useful satellites become to one side, the more attractive they will be as military targets to the other. That fact seems to be moving space from a behind the lines support sector into a combat arena.

# CHAPTER SIX

# ASAT: Weapons in Space

Since the mid-seventies The Aerospace Corporation has run a space warfare simulation on its IBM 370/155 computer. The game is a complex computer program called the Satellite Engagement Scenarios Model. Whizzing through the imaginary space of the SES Model are 112 satellites, divided – as U.S. war games usually have it – into Red and Blue teams.

Just as real satellites are elements in larger military systems (including other satellites and supporting ground stations), the SES satellites are also tied into hypothetical systems. The computer games explore how these opposed systems interact with each other, which satellites or systems are vulnerable to attack from those on the other side and under what conditions. The human operators feed in the conditions they want to set for the imaginary space battles through the 'Star Wars controller program segment.'

In 1977 the Martin Marietta Corporation designed some additions to the Satellite Engagement Scenarios Model. The new variations in the program included an electronic warfare model to study the effects of enemy radio interference with communications between ground stations and satellites. The model added two new systems to the Red forces – ground jammers and space jammers – and one new set of Blue ground communications stations. The model already included Red ground-based anti-satellite lasers and Red space lasers, but the Martin Marietta additions added possible laser countermeasures. A company report on the new programs said, 'For each blue system a curve of the amount of sensor damage versus the distance of laser attackers (unique relative to laser type) will be added to the input.'

The SES Model has been running since the mid-seventies, and its simulated world is well ahead of current reality. So far, nobody has fought space battles and the American and Soviet abilities to do so are limited. But both sides are working to make war in space a genuine possibility.

The first test of an ASAT (anti-satellite) weapon was an accident. In October 1962, the Atomic Energy Commission and the Air Force carried out a high-altitude nuclear weapon test code-named STARFISH. STARFISH, as it turned out, fatally damaged a number of orbiting satellites. What's more, not one of them was in line of sight of the nuclear detonation. They were all wrecked by the massive doses of high-energy electrons that the weapon had suddenly injected into their paths. Whether related to lessons learned from the STARFISH test or not, the world's first anti-satellite weapons were nuclear-armed. The United States had been studying the idea of an inspection and interception vehicle, under project SAINT, but that was canceled.

Responding to the possibility that the Soviet Union would deploy an orbital bomb, President Kennedy early in 1963 approved plans for developing 'an active anti-satellite capability.' By 1964 the Army had tested and deployed a few of its test anti-ballistic missiles (the Nike-Zeus) as an 'interim' ASAT capability. The Army shut down that system in 1967. But in March 1964, the Air Force had begun a series of tests from Johnston Island, southwest of Hawaii, that took on the code name SQUANTO TERROR.

The more mundane name for the whole project was Program 437. In the SQUANTO TERROR tests, a Thor missile lofted a simulated nuclear warhead up to 700 nautical miles high and up to 1,500 nautical miles down range. The targets were dead U.S. satellites or pieces of debris from U.S. launches. Since the 1963 Limited Test Ban Treaty forbade nuclear tests above ground, the test rockets carried a dummy warhead. The Air Defense Command judged the success of the tests by how close they came to the targets. With a nuclear weapon, anything within 5 nautical miles would have been close enough. The very first test was a success. By 1965, some shots were getting as close to the target as nine tenths of a nautical mile. After the first three tests in 1964, the Air Force declared the system operational, capable of attacking two satellites a day – if, of course, they passed within range of the Johnston Island base.

Between 1964 and 1968 the Air Force ran as many as sixteen SQUANTO TERROR exercises. Despite the 1967 treaty banning nuclear weapons in orbit, it kept the Program 437 launchers ready until 1975. Not only had the threat of orbital bombs not materialized, but the Air Force must have realized that as an anti-satellite weapon its system had the definite disadvantage illustrated back in 1962 by the STARFISH test: a nuclear explosion in near-earth space was likely to do as much damage to U.S. satellites as it was to Soviet target spacecraft.

In 1968 the Soviet Union began testing its own ASAT weapons. From then until 1982 it had conducted twenty tests of interceptor satellites against target satellites. By 1980, the Joint Chiefs of Staff were declaring the Soviet ASAT system 'operational.' In their tests, the Soviets first launch a target satellite at an inclination of 65 or 66 degrees to the equator (in the earliest tests, the inclination was between 62 and 64 degrees). The targets have been in orbits with altitudes ranging from about 340 to 1,200 miles, but in the most recent tests they have been in circular orbits about 600 miles up. Days, or even weeks, after launching the targets, the Soviets launch an interceptor satellite. In some earlier tests the interceptor more or less matched the orbit of the target, but in the latest tests the interceptor stays in a lower orbit, then increases its apogee (highest altitude) to approach the target. In this 'pop-up' maneuver, the interceptor's orbit stays highly elliptical, so it passes by the target at around 400 miles per hour.

The interceptor would 'kill' its target by exploding into a cloud of high-speed metal fragments. In actual tests, one or two interceptors may actually have exploded near the target. In others, the Soviets commanded the interceptor to a higher apogee, then exploded it. In still others, they have ordered it to re-enter the atmosphere and burn up.

Private satellite watchers using publicly available data

have been able to make some deductions about the Soviet ASAT tests, but the most detailed information comes from Air Force leaks to *Aviation Week and Space Technology* magazine. That's how we learn, for example, whether the Aerospace Defense Command judges a Soviet test to be a 'possible success' or not. In the tests conducted through 1980, the Air Force judged them possible successes if the interceptor passed within 1 kilometer (.62 miles) of the target: ten of the first seventeen met this standard. Then, in 1981, *Aviation Week* reported without comment that the Air Force now considered *8* kilometers – 4.8 miles – close enough.

It's clear from available information that the Soviet 'killer satellite' is a weapon of limited capabilities. The Soviets don't seem to have settled on a target-finding device – the last one in 1981 used radar, but the preceding four (of which only one was considered a possible success) used some kind of infrared sensor. Anonymous American intelligence sources have called the guidance system 'not very impressive.' The interceptor gets into space aboard a satellite-launching version of the Soviets big SS-9 ICBM. It's a massive, liquid-fueled vehicle, and even if, as some assert, they can get one onto the pad and ready to launch with ninety minutes' notice, the Soviets probably can't launch very many of them in a short period of time.

Theoretically, the killer satellite should work against any satellite at altitudes under 1,200 miles, but the fact that it has only been tested at inclinations from 62 to 66 degrees indicates some kind of limitation. Within that range of inclination, about the only U.S. military satellite the Soviets might go after is the Navy's ocean reconnaissance satellite, which is far from the most important. The more important Satellite Data System strategic communications satellite does fly at an inclination of about 63 degrees, but at its perigee (lower altitude), it swoops by the earth at about 1,000 miles an hour faster than the Soviet interceptor.

Assuming the Soviets could use their ASAT to attack U.S. satellites going over the poles, they could also go after photo reconnaissance satellites, weather satellites, and the Transit type of navigation satellite. What they could not go after are the new Global Positioning System spacecraft and the important geosynchronous satellites for missile attack warning, electronic intelligence, and communications. Still, if they put it on some other, bigger rocket booster, they might get the interceptor into the higher orbits. And if they keep working at it, they're bound to build a better ASAT. Someday, they may build one as good as the one the U.S. Air Force is about to start testing.

The new Air Force ASAT probably reaches a speed of about 2 miles per second, or 7,200 miles an hour. An earth satellite in a low circular orbit, around 100 miles up, goes about 17,000 miles an hour. A Molniya-orbit satellite – one following the highly elliptical path typical of the Soviet Molniya communications satellites – reaches more than 23,300 miles an hour at its perigee, which is around 200 miles up. The new U.S. weapon does not have to match these speeds because it simply cruises into the path of the target satellite. The target virtually destroys itself by crashing into the U.S. ASAT at a relative speed of 10,000 miles an hour or more.

For the time being this weapon goes by the unassuming name of PMALS – the Prototype Miniature Air-Launched System. The PMALS payload, the computerized kamikaze that is rammed into the target satellite, is a little gem of technical ingenuity called, in the officially 'sanitized' documents, the Miniature (Deleted) Vehicle. The (Deleted), the trade journals tell us, censors out 'Homing Intercept.' The Miniature Homing Intercept Vehicle, the MHIV, is indeed miniature: a cylinder about 12 by 13 inches jam-packed with state-of-the-art gizmos. Looking out from the center of the cylinder are eight small telescopes. Their job is to gather infrared light from the target satellite and focus it on an electronic sensor in the focal plane (the sensor is cooled to an extremely low temperature so that it can register the faint infrared emission from the target).

When the satellite is free of its carrier rocket, it is already on a near-collision course with the target satellite. What it needs to do to make sure the target crashes into it is to maneuver from side to side or up and down according to the direction of the target registered by the telescopes. The maneuvering power comes from fifty-six small rocket tubes forming the outer shell of the cylinder. These rockets exhaust out of a ring of fifty-six corresponding holes around the center of the vehicle, so their propulsive force is at right angles to the direction of travel.

Steering the miniature vehicle into the exact path of a dim object approaching at 10,000 miles an hour sounds difficult enough, but there's more. When the MHIV leaves the carrier rocket, not only is it traveling at enormous velocity, but it's spinning twenty times a second. The spin is necessary to stabilize the vehicle and keep its telescopes pointed in the right general direction. What's more, each of the fifty-six solid-fueled rockets fires only once; that means that the timing of the maneuvering bursts has to correlate not only with the calculated position of the target, but also with the exact position of the rocket tubes as the vehicle spins. This complex maneuvering task forced the Vought Corporation engineers building the weapon to design a whole new guidance system for it. That guidance system incorporates not only the infrared target trackers but a recently invented laser gyroscope to keep track of the vehicle's revolutions.

Traditional gyros use the momentum of a spinning metal wheel to keep track of changes in motion: movements against the plane of the spin register as torque on the wheel's axis. But the ring laser gyro has no moving parts. Instead, a split laser beam travels in both directions around a ring of mirrors. Turning the ring in one direction or the other measurably desynchronizes the two halves of the beam. The ring laser gyro inside the MHIV keeps exact track of spins and tells the weapon's electronic brain when the rockets are in the right position to fire.

Vought's design for the MHIV comes out of research in ABM (anti-ballistic missile) weapons originally sponsored by the Army. But the Air Force's method for getting the little projectile into the general vicinity of the target really turns it into an entirely different weapons system. Where the proposed ABM system would have one larger but slower rocket carry a package of several of the miniature vehicles into the path of an oncoming wave of Soviet nuclear warheads, the PMALS uses a single two-stage rocket, launched from an airplane, to deliver each MHIV into space.

The two-stage rocket, which is what hurls the little vehicle into the path of the target, consists of two more or less 'off the shelf' rocket designs. The first-stage rocket engine comes from the Boeing-built Short Range Attack Missile (the SRAM is an air-to-ground missile that Air Force B-52 and FB-111 bombers carry; in a nuclear war, the SRAM would blast Soviet defenses in the path of the bombers). The second stage of the PMALS comes from Vought's Altair Three, which is also the upper stage sometimes used on the Vought Scout space-launch rocket. The Altair stage on the anti-satellite weapon comes with a 'spin table,' which starts the miniature vehicle rotating twenty times per second.

The whole package – the two-stage rocket plus the MHIV – fits together as a missile a little less than 18 feet long, about 20 inches in diameter, and weighing about 2,600 pounds. That missile hangs on a special pylon (built by Boeing Aerospace) attached to the underbelly of a McDonnell Douglas F-15 jet fighter. The fighter really becomes the first launching stage of the whole system. So McDonnell Douglas has the contract from the Air Force for modifying the F-15 to launch the missile. Besides fitting out the aircraft with attachments for carrying the missile, the company has to adjust the plane's electronic guidance systems so they can communicate with the missile and help the pilot launch it accurately.

Boeing Aerospace has the job of making sure that all the parts of the system – the Vought MHIV and second stage, the Boeing first stage, the connecting pylon, and the F-15 – fit together and work right ('total system integration'). That includes a special kit that will make it possible for Air Force ground technicians to convert virtually any F-15, stationed anywhere in the world, into an anti-satellite weapon carrier in about six hours. The Air Force Space Command plans to station the first kits, and therefore the first anti-satellite F-15 squadrons, at Langley Air Force

Base (near Hampton, Virginia) and McChord Air Force
Base (near Tacoma, Washington) in 1985.

But the headquarters for the whole system will be back
inside Cheyenne Mountain, at SPADOC – the Space Defense Operations Center. Boeing also has the contract to
develop the anti-satellite mission control center at SPA-
DOC. Officers there will tell the F-15 pilots where to aim
the missile to get the MHIV in the general vicinity of the
target satellite so that it can maneuver in for the deadly
collision. When the Air Force tests the system in 1984, it
will aim the missiles at special target satellites, built by the
Avco Corporation. But of course the ultimate targets will
be Soviet military satellites. So the Space Defense Operations Center commanders will be relying on the SPA-
DATS – the Space Detection and Tracking System – to
supply the information needed to direct anti-satellite
attacks.

With the prospect of space wars on the horizon, the Air
Force has been tooling up the SPADATS to gather better
data faster. For example, the Electronic Systems Division
has set up a 'Pacific barrier' of radar stations to fill a gap
that the Hawaiian and Aleutian islands stations miss. One
of the radar sets used to be based in Thailand, then went
into storage in the Philippines, and now is up and working
there. A new system went to Guam. And at the Kwajalein
missile test range in the Marshall Islands, the Air Force
converted a radar that used to track incoming test missile
warheads into a full-time satellite watcher.

Another new tracking system uses telescopes rather than
radar. For many years the Air Force has maintained a
globe-girdling belt of optical trackers called Baker-Nunn
cameras. The Baker-Nunn cameras, like astronomer's
telescopes, are sidereally mounted: a swiveling platform
keeps their line of sight moving opposite to the earth's
rotation. The result is that the stars appear to stand still in
the night sky. But examination of the photographs taken by
the Baker-Nunn shows some objects that don't stand still;
instead, they show up as streaks of light against the fixed
background of stars. Those streaks are the earth satellites
the SPADATS analysts are looking for. With cameras like

these, on a clear night a satellite the size of a basketball will show up at 20,000 miles.

Every night the Baker-Nunn cameras devour hundreds of feet of film, and an hour and a half can pass between the time the picture was taken and the processing and analysis of the film. But now TRW and several subcontractors are supplying a new system, the GEODSS (Ground-Based Electro-Optical Deep Space Surveillance), that works a hundred times faster. Instead of registering on film, the pictures go through a TV camera. And with a computerized method of analysis, two operators at a GEODSS station can almost immediately spot the satellites, determine their orbital characteristics, and compare the information with previous observations. The first GEODSS station opened up for business at the White Sands Missile Range, New Mexico, in 1981. Four more are planned: one in Korea, one in Hawaii, and two at unannounced sites in the eastern Atlantic and the Middle East.

Not satisfied even with the new speed of the GEODSS, the Defense Advanced Research Projects Agency has been working with the Air Force to develop an even more advanced system that would replace the GEODSS TV tubes (which scan the telescope image with an electron gun) with mosaic detector arrays similar to the sensors being developed for satellite-based reconnaissance. The new sensor, code-named TEAL AMBER, would, like its space-based cousins, 'stare' at the target instead of scan it; the sensor would work faster and detect even fainter objects than the GEODSS television system.

The space watchers want to know not only what the Soviet satellites' orbits are, but what the satellites themselves look like. At Kwajalein, a DARPA radar system can form images of satellites. In Lexington, Massachusetts, the 'Haystack' radar, run for the Defense Department by the MIT Lincoln Laboratory, does the same thing. The Aerospace Defence Command's MOTIF (Maui Optical Tracking and Identification Facility) in Hawaii watches satellites with a laser beam radar (bounces light instead of radio waves off the target) and a 'compensated imaging' telescope. With compensated imaging, the telescope mirror is subtly distorted to compensate for the twinkling of images caused by the atmosphere. (The Defense Advanced Research Projects Agency is exploring the same technology for us on space-based surveillance systems and for focusing the beams in space laser weapons.*)

The trouble with ground-based radar systems is that they can only track satellites out to about 2,000 or 3,000 miles. The trouble with ground-based optical systems is that they work only at night and with clear skies. The trouble with both kinds of systems is that they have to be based in foreign countries to cover all the skies, and it's getting harder and harder for the United States to get foreign countries to accept U.S. bases. These shortcomings in the present SPADATS system have piqued Air Force interest in space surveillance systems to be based on satellites. One research project in the works for some years now has been SIRE, the Space Infrared Experiment. Originally, the Air Force wanted to put SIRE on a free-flying satellite for long-term testing; then it decided it could ride on a space shuttle 'sortie' mission, in which the Air Force astronauts would conduct the experiment right on board the shuttle orbiter. Most recently, the Air Force seems to be leaning again toward a separate satellite.

The computers at SPADOC, inside Cheyenne Mountain, are already keeping track of the 4,500 or so artificial satellites, ours and theirs. But as SPADOC moves into the 'warfighting' mode, and as its new global network of sensors gathers more timely information, it's going to need better and faster computer programs and display consoles. Ford Aerospace and Martin Marietta have been carrying out 'definition' studies for the equipping of SPADOC Phase 4; one of the companies will get the final $85 million contract. The contractor is supposed to finish its work by 1985.

---

* The Pentagon has one more way of getting pictures of Soviet satellites. During the first flight of the space shuttle, some of the heat-resistant covering tiles fell off during launch. NASA got the Pentagon to take a picture of the orbiter *Columbia* from the KH-11 spy satellite. This of course only works when the target satellite happens to pass more or less directly beneath the 170-to-300-mile-high KH-11 (and Soviet reconnaissance satellites and manned space stations are about the only ones that regularly do), but the Air Force must have a few pretty good close-ups of some Soviet spacecraft.

Defense officials are ambiguous about whether the SPA-DOC and the ASAT weapons under its control are supposed to be more for defense or for attack. The usual justification for the U.S. anti-satellite program is the existence of the 'operational' Soviet interceptor satellite. The idea is that if the U.S. can threaten to retaliate against them in kind, the Soviets will be less likely to use their weapon. But in the spring of 1981 Brigadier General Ralph Jacobson, then the Air Force research chief for space systems, reversed the usual emphasis. He told a subcommittee of the House Appropriations Committee: 'The ability of the Soviet Union to use military power on a worldwide basis is increasingly dependent on effective and reliable operation of various satellite systems. These systems enhance the performance of Soviet surface, sea and aerospace forces and represent a major threat to U.S. and Allied sea, ground and aerospace forces. Thus, the U.S. has a legitimate military need for an ASAT capability to remove the current sanctuary status the Soviets enjoy in space. In addition, posing a threat to Soviet satellites may help deter Soviet use of their operational ASAT capability.'

In 1979 the Pentagon's deputy chief of research for strategic and space systems, Seymour Zeiberg, told another committee that it was '. . . important not to couple our anti-satellite program with the Soviet's anti-satellite program. The principal motivation for our anti-satellite program is to put us in a position to negate Soviet satellites that control Soviet weapons that could attack our fleet. That differs, in my mind at least, from a consideration that if they have one we ought to have one and we can develop some deterrence in the use of anti-satellite systems.' The satellites Zeiberg referred to are the Soviet ocean reconnaissance satellites, which might tell the Soviet Backfire bombers where to aim their missiles and bombs to strike U.S. aircraft carriers.

So the main task of the SPADOC officers in charge of the Miniature Air-Launched System will be to spot likely targets among Soviet military satellites. A more challenging task, but one that the new and improved SPADOC is planning to take on, will be to spot a suspected Soviet interceptor satellite within minutes after it leaves the ground, determine whether it is about to attack an American satellite, scramble one of those anti-satellite-equipped F-15s, and guide the U.S. weapon in a counterattack. All of this would probably have to happen within about forty-five minutes. If it can be done at all, it can probably be done only when crisis conditions have led the Aerospace Defense Command to be expecting an attack and only when the attack is on certain satellites in certain orbits.

The Miniature Vehicle might work as a defender if it could ride piggyback on the satellite it is supposed to defend. It would have to spot and lock on to the attacking interceptor at a long enough range to prevent damage to the protected satellite. But the Air Force space planners recognize that defensive weapons will have their limitations against even the present, limited Soviet anti-satellite weapons and they will be even less useful against other threats on the horizon. They're now devoting more and more research and study to satellite attack measures and countermeasures, active and passive methods for making satellites more survivable against the imagined threats of the future.

If the SPADOC gets warning of probable attack on a particular U.S. satellite, the best defense may be to just get out of the way. The Soviets have to adjust the orbit of their killer satellite so that somewhere the killer intersects the orbit of the target. If the orbit of the target changes at the last moment, the interceptor will miss. So if SPADOC can tell the target satellite's operators that their bird may be under attack, they can maneuver it higher or lower and evade damage. For example, the Big Bird spy satellite is a low-orbit sitting duck for an ASAT. But it also has maneuvering rockets to change and maintain its orbit. It just might escape the clumsy Soviet interceptor satellite. With this sort of tactic in mind, SPADOC is getting all the military satellite operators (and trying to get all the U.S. civilian satellite operators) to keep it constantly informed of the status of their spacecraft.

The Air Force Space Division is supporting research (with unnamed contractors) that would eliminate the SPA-

DOC middleman. The idea is to put sensors right on the spacecraft that would detect if it were being tracked by radar or laser radar so that its own computer could direct evasive maneuvers. Of course if the attacker is using an electrooptical tracking device – like the infrared telescope set on the Miniature Homing Intercept Vehicle – the detect-and-evade scheme wouldn't work.

If the United States and the Soviet Union get serious about going after each other's satellites, the potential threats are considerable. From the Soviet point of view, the U.S. Miniature Air-Launched System probably already represents an escalation in the space arms race: if it works as designed, it will be a much more formidable weapon than the Soviet killer satellite. (In fact, as a backup, in case anything goes wrong in the development of the Miniature Vehicle, the Air Force is working on a simpler system, more like the Soviet one.) The air-launched weapon will attack from any direction, making defenses more difficult. Unlike the Soviet ASAT, which goes into at least a partial orbit, and sometimes two orbits, before it strikes, the U.S. system is a 'direct ascent' interceptor. As a small missile launched from a relatively small airplane, it will give virtually no warning to Soviet radars or warning satellites that an attack is imminent.

The U.S. system, unlike the Soviet system, needs no specialized missile launching pad: if the Air Force wanted to it could launch many ASATs from many locations in a very short period of time – and much more cheaply. As carried on the F-15, the U.S. weapon will be limited, like the Soviet one, to low-altitude targets. But mounted on a ground-based rocket booster, the American weapon could reach the high-altitude geosynchronous satellites as well. Mounted on a much larger booster, so could the Soviet interceptor, but again, because it's so big and heavy and slow, it would lack the flexibility of the Miniature Vehicle.

A nonnuclear interceptor attack is just one way the space powers might ultimately try to interfere with each other's orbiting military assets. The Pentagon communications, command, and control planners, for example, take seriously the possibility that high-altitude nuclear bursts could put

U.S. satellites out of commission. Intense X rays from the bomb can produce what is called System Generated Electromagnetic Pulse in a satellite's electronics. The circuits overload and burn out. The Defense Nuclear Agency has tested one set of satellite circuits in an underground nuclear test, and plans more. The purpose of these and other experiments is to find out how much the satellites can take and to design electronics that are 'hardened' against electromagnetic pulse effects.

Explaining these efforts, in 1980 the Defense research chief, William Perry, said, 'It is not possible at present to prevent the destruction of a satellite by a dedicated direct attack. The goal of satellite survivability programs is to prevent the simultaneous loss of many satellites from a single detonation. Such a threat could occur from a high-altitude test, an Nth country event, an ABM detonation, or an overt act by an aggressor to sterilize space and render a significant number of critical systems inoperative prior to some hostile act or political threat.' He suggested, for example, that the Global Positioning System satellites could be '. . . irradiated and placed at risk by a few properly placed detonations.'

For Perry the subject was full of uncertainties, and that was why the tests were necessary. It's such an uncertain subject, in fact, that there seems to be disagreement among the experts as to how significant it is. In 1982, the head of the Defense Advanced Research Projects Agency, Robert Cooper, said, 'I don't think we have the right ideas yet about how to go about assuring survivability, and I think we have a preoccupation with hardening against nuclear radiation that is totally out of bounds and unnecessary.'

Nuclear bursts still carry the risk of damaging as many friendly satellites as enemy ones. More subtle means of electromagnetic interference are conceivable. One would be to interfere with communications between the satellite and the ground. This applies not just to the militarily useful information the satellite may be relaying or sending, but also to the 'telemetry' channels – particularly the flow of ground-base instructions to the satellite. At some expense, radio signals can be encrypted and made resistant to jamming. Already in the more traditional arenas of combat – land, sea, and air – the United States and the Soviet Union compete in an endless shadow duel of electronic warfare weapons (EW), electronic countermeasures (ECM), electronic counter-countermeasures (ECCM), and so on.

There just isn't any publicly available information about what the Soviets are up to in the way of space-related electronic warfare technology, but U.S. military satellite designers are clearly trying to prepare for the worst. Not only are they trying to securely encode and make jam-resistant the communications between satellites and the ground, but they're trying to make the satellites less dependent altogether on ground commands. Hence Space Division commander General Henry's concern with building new satellites that can navigate and 'keep house' with their own autonomous on-board computers.

A cruder form of electromagnetic interference, which the Soviets may or may not be working on, would be the radio frequency pulse weapon. Like the electromagnetic pulse effect induced by nuclear weapons, the radio frequency pulse weapon would try to damage the satellite's electronic circuits, but instead of X rays, the radio wavelengths that the satellite's antennae were sensitive to would induce the overload.

Yet another kind of anti-satellite weapon borrows an idea from naval warfare:) the mine. A space mine would be a satellite 'parked' not far from its target, following it around in its orbit. The space mine would have its greatest potential against the warning, intelligence, and communications satellites in the geo-stationary orbit. Not only is the 'parking' job easier, but the orbit is a relatively small corridor in space where every space-faring nation has a legitimate interest in keeping its satellites. A mine can easily be stored in one slot in the lane and then moved over to do its damage in a short time. Whether it's parked near the target or stored and moved later, it might be disguised as something else – say as a communications satellite. The target's owner might not know whether the foreign object was really an explosive mine until it was too late. The mine would be somewhat less of a problem in the other kinds of orbits, because a satellite shadowing another could prima facie be suspected as hostile. The target's owner might then try to maneuver it out of the way or, at least before war had broken out, take up the grievance with the suspect's owner.

In the fall of 1981 the editors of *Aviation Week and Space Technology* reported an intelligence leak – of dubious origin – that said the Soviets had launched a strange new anti-satellite weapon launcher into orbit. The story was that the satellite Cosmos 1267, which had docked with the Soviet Salyut 6 space station, was '. . . equipped with firing ports to eject 1-meter-long miniature vehicles guided by infrared sensors. . . . Docking of this anti-satellite weapon platform with Salyut 6 means the USSR would be able to use a manned Salyut 6 to direct anti-satellite attacks against U.S. spacecraft or to protect Soviet satellites against a U.S. retaliatory attack.' Cosmos 1267 was a type of satellite usually used to resupply the manned station.

It's extremely doubtful that the Soviets have the technology to locate U.S. targets and command anti-satellite weapons from space. It's less likely that they would base anti-satellite weapons on a costly manned space station, exposing it needlessly to risk of counterattack. The *Aviation Week* space editor later said he himself didn't believe the story 'at first.' He didn't say what, if anything, caused him to believe it later. And the magazine never published any more information that would give the farfetched tale any credibility. In April 1982, the Soviets launched their next space station, Salyut 7, and abandoned Salyut 6 altogether. Later, it re-entered and burned up in the atmosphere.

One thing is certain: a '1-meter-long' rocket in orbit wouldn't get far from the mother ship. The *Aviation Week* editors later tacitly admitted the implausibility of the Cosmos 1267 'threat' when they speculated that its purported missiles might be used for close-in defense of the mother ship or that somehow it might later be put into geosynchronous orbit, where the 'delta vee' from one orbital slot to another is quite low. In the case of attack by the 45,000-foot-per-second U.S. Miniature Air-Launched System, the

Salyut space station with or without a docked 'battle station' would probably never even see the little super-bullet coming.

The first laser was built in 1960, and it wasn't long after that military men started thinking about turning it into a weapon. 'Laser' is an acronym for *l*ight *a*mplification by *s*timulated *e*mission of *r*adiation. Without going into the quantum mechanics of it, a laser is an intense beam of light of a single color or wavelength. The intensity is especially great because the light waves are 'coherent' – in lockstep with one another. The advantage of this coherence is that the light can be much more efficiently focused than light from ordinary sources. In fact, the wavelength of infrared light emitted by a carbon dioxide gas laser can be focused about 100 *million* times more intensely than the light from an ordinary light source of the same size.

In most kinds of laser generator, molecules of one kind or another (carbon dioxide is one example) are 'pumped up' by some external source of energy into a higher 'energy state.' Then, as the electrons of those molecules return to a normal state, they emit photons – tiny packets of light – that are all of the same amount of energy and electromagnetic frequency. A carbon dioxide laser device, for example, emits a particular wavelength of infrared light. This pumping and emitting process takes place in a chamber called the laser 'cavity,' where the photons bounce back and forth off mirrors. As they collide with the molecules in the cavity, they produce more and more of their own kind. In the middle of the cavity is a 'scraper' mirror, a mirror with a hole in the center. The reinforcing photons move back and forth through the hole, but the scraper sends some of them out through a hole in the side of the cavity where they can be focused into a beam by yet another mirror. Because the light is also 'coherent,' the beam can be exceptionally narrow and intense, with very little spreading even at great distances.

If the beam is powerful enough, it may be useful as a weapon. Since at least 1962, military men have been talking about the potential of lasers as instruments of war, but more than $2 billion of research has yet to produce a really practical field weapon. There's now a scientific and technical debate on in the U.S. military-industrial community about just how soon, if ever, various kinds of laser weapons are going to be feasible.

Meanwhile, the satellite survivability thinkers are figuring out how to counter the hypothetical laser weapons. Because satellites often rely on quite delicate instruments either to do their jobs or just to 'keep station,' a laser weapon wouldn't necessarily have to be super-powerful to do them harm. A laser based on the ground, even though the atmosphere would degrade the intensity of its beam, might damage the optical horizon sensors that keep some satellites properly oriented, or might overload the light detectors in a spy satellite's scanning telescope.

In 1975, reports leaked from the Pentagon that the Soviets had temporarily 'blinded' a U.S. early warning satellite by aiming a ground-based laser at its infrared sensors. Later, the generally accepted explanation was that gas-field fires in the Soviet Union had produced the blinding effect. The next major leak on the subject of a Soviet ground-based laser came in 1980: a CIA estimate given to President Carter reportedly said that the Soviets had recently deployed such a weapon. According to the Associated Press, though, Pentagon officials were skeptical, saying, '. . . it's a possibility, but no more than that.' The editors of *Aviation Week*, on the other hand, passed on the leak as gospel: 'The USSR already has operational a high-energy laser weapon capable of destroying U.S. satellites in low earth orbit.'

A ground-based laser, in any case, is a limited weapon against satellites. Low-flying reconnaissance satellites are about the only suitable targets, and then only if the spacecraft happens to be flying over the right spot on a cloudless day. Besides, it is possible to build countermeasures into the sensitive optical devices. For example, there are spectral filters that can reject the laser's wavelength, or at least engage a shield to protect the sensors while the laser is aimed at them. Colonel Donald Henderson, Air Force Space Division expert on anti-satellite warfare, recently told a meeting of the American Astronautical Society that 'laser hardening and sensor protection technology is nearly perfected for defensive applications.'

Every once in a while an intelligence leak appears to the effect that the Soviets are developing a space-based laser weapon for anti-satellite purposes. A laser in space has the advantage that there is no atmosphere to distort and dissipate the power of its beam. A high-energy laser in space might actually burn or punch a hole in a satellite, not just damage its sensors. Getting a working laser weapon into space will be no easy feat of engineering. There is much debate among military and intelligence analysts and scientists as to when, if at all, either the United States or the Soviet Union can do it.

Lasers in space for anti-satellite purposes might not necessarily be the ultimate anti-satellite weapon. One possibility is that both sides would put such weapons up, setting the stage for a laser duel of uncertain outcome. Short of that sort of star war, though, the satellite defense planners have less dramatic means available for countering the laser threat. For example, if you can build a mirror strong enough to aim at the laser beam in the first place, you should be able to build one strong enough to fend off an enemy beam. A defensive mirror even of low quality might protect the satellite; a higher-quality one might even reflect the beam back on the enemy weapon. Scientists are also developing 'ablative' paints and structural materials that can erode away under a laser beam, leaving the equipment under the skin unharmed.

There are uncertainties, then, about how effective lasers could be against satellites. But among the tasks imagined for space laser weapons – destroying ballistic missiles, destroying bombers, or destroying satellites – the anti-satellite task would be the easiest. Defense officials have said that if they were to deploy a laser weapon in the next few years, attacking satellites would be the most they could expect to do with it. And they judge that if that's all you can do with it, it's not worth doing. George Millburn, who in 1981 was the acting Number Two Pentagon official for research, said, 'The anti-satellite scenario of importance to laser weapons is one requiring very rapid negation of many enemy satellites. Although such situations can be imagined, none of these anti-satellite scenarios seemed appropriate.'

A better candidate for an anti-satellite laser platform might be a high-flying aircraft. At 30,000 or 40,000 feet, the densest part of the atmosphere would lie below. If the plane fired a powerful laser upward at a satellite, enough energy might well get through to cause damage. Since the airplane flies above the clouds, the weapon doesn't need a clear day to work. What's more, the airplane is mobile – you can move it to where the satellite will be passing over, instead of waiting for it to come along in its own time. The Air Force has already put a modest laser test weapon on a C-135 (military version of the Boeing 707) 'test bed,' but it's had only partial success in shooting down even close-in target missiles.

The most enthusiastic laser proponents see space lasers as the ultimate anti-ballistic missile weapon. Their arguments are considered in Chapter Eight of this book. Nearer at hand, though, is the accelerating competition in anti-satellite weapons and measures for dealing with those weapons. The next chapter looks at ways we might cope with war in space.

# CHAPTER SEVEN

# War in Space?

To hit a satellite with a 'killer satellite,' a missile, a mine, or a laser, the attacker has to find it first. The Air Force is looking for defensive countermeasures that would make U.S. military satellites harder to find. One such countermeasure is maneuver – if attack seems imminent, move the target out of the way. By the time the attacking weapon arrives at the point in space where it was aimed, the target will then be somewhere else. Continued random movements might keep the attackers from ever zeroing in on the target at all.

Other countermeasures are designed to make the target less visible. These techniques are similar to those that will go into the Stealth strategic bomber: the idea is to minimize the 'signatures' by which the spacecraft might be spotted or identified. There are, for example, paints or coatings that do not reflect radar waves well, making for a smaller blip on the enemy's screens. There are also 'deception jammers' – radar emitters right on the satellite that fool the attacker's radar receivers into misestimating the position or distance of their target.

If you can't hide from the attacker, you may be able to confuse him. Decoys, built with the radar, visible light, and infrared signatures of the real thing, could force the attacker to dissipate his forces on false targets. Or the real satellite itself might in a sense become the decoy. If there were 'silent' spare satellites stored in high orbits, they could be brought down to replace the destroyed primary satellite. This is one idea the designers of the new Milstar communications satellite system are looking into.

Yet another defensive measure would be to dilute the effects of an attack on any particular satellite by building lots of redundancy into the whole system. The GPS navigation satellite system, for example, will work with eighteen satellites (and it could be many more). If the services of some of the satellites should be lost, the system planners believe that the system as a whole will 'degrade gracefully.' By that, they mean that the service it provides will be less complete and the accuracy reduced, but it will still be there – a few successful attacks can't wipe out the whole system.

In contrast, an attack (not feasible in the near future) on the three U.S. missile early warning satellites (Defense Support Program) and perhaps one or two spares would pretty much wreck the whole satellite early warning system. Ball Aerospace Corporation in Boulder, Colorado, has suggested that it could build a large number of small missile-detecting satellites which, combined with many decoys, would form a warning system almost impossible for the Soviets to destroy. The Ball engineers think they could launch the whole system in one space shuttle trip.

Still another technique for neutralizing the effects of an anti-satellite attack would be to launch replacements for the destroyed satellites. In Pentagon studies, this idea goes by the name of 'reconstitutable satellite systems.' The scenario calling for reconstitution of satellite systems is usually nuclear war. In this scenario, the planners assume that the major launch centers at Vandenberg and Kennedy Space Center have been destroyed. But some ballistic missiles, such as Trident missiles hidden at sea in submarines, might be able to launch small communications satellites into relatively low orbits. If the major geosynchronous communications satellites had been destroyed, these small satellites might restore some connection between surviving military commanders and their submarines and bombers.

Colonel Stephen McElroy, space launch research and development chief at Air Force headquarters in the Pentagon, explains his unofficial ideas about how such an 'advanced space launch capability' might supplement the space shuttle. 'During peacetime we require some "X" amount of space assets to do our missions. I submit . . . that we don't really need all that "X" in a nonpeacetime environment. For all intents and purposes, with space systems we're really at peace up until total nuclear war. I think that up until the point where we have a nuclear interchange, it's reasonable to think that our satellites are in sanctuary to a degree that we can depend on most of our peacetime assets to do the mission for us. I think that's true even up to theater nuclear war, just because the Russians need their satellites as bad as we need ours. So we use the shuttle to get up those regular assets for the peacetime missions, and then have an ability to launch the smaller payloads in the post- or trans-attack [nuclear] environment.'

Colonel Charles Heimach, who has worked for the Air Force and the Defense Advanced Research Projects Agency on satellite attack and defense for more than twenty years, has different ideas. He thinks that after full-scale nuclear war has started (or, as he puts it, 'post-SIOP' – after execution of the Single Integrated Operational Plan), the satellite defense job will be easier than before. His point is that sophisticated anti-satellite systems require a lot of tracking and command support from the ground – and the U.S. could eliminate the Soviet ground stations in a nuclear war. Even if the Soviets managed to launch anti-satellite weapons at the very beginning of the nuclear exchange, the U.S. could order its high-altitude satellites to maneuver out of the way within minutes of the Soviet missile launches. By the time the anti-satellite weapons arrived where the U.S. satellites had been, those targets would be gone. And the Soviets would no longer have the ground tracking stations to find them again or the communications links to update the anti-satellite weapons' guidance systems.

Unlike McElroy, Heimach does not dismiss the possibility that the Soviets would make limited attacks on U.S. satellites before full-scale nuclear war broke out. But he has a theory about how to deter them from doing it: the principle is to cause '. . . the Soviet decision-maker to reject the idea of space attack because of the risk that may result from a U.S. reaction thus placing the Soviets at a disadvantage.' More specifically, he envisions making it necessary for the Soviets to launch a 'massive space attack' on U.S. warning and communications satellites if they hope to have any significant military effect. The Soviets would be taking the risk that such a massive onslaught would signal the U.S. that nuclear war was imminent and mobilize its forces accordingly, if not actually launch a preemptive attack itself.

If the U.S. put up many smaller, simpler satellites to do the jobs now done by big ones, Soviets would have to launch a massive attack on a large number of satellites to keep them from carrying out their mission. The U.S. could further complicate the Soviet problem by designing its satellites so as to hide just exactly what their mission was – say, making warning satellites look like communications satellites and vice versa.

Heimach is pessimistic about the Pentagon actually carrying out the measures he thinks necessary to improve the chances of survival of U.S. military satellite systems. Like other spacemen, he sees resistance to unified budgets and organizational arrangements for space activities. He said, 'The problem is, we have been unable to articulate the function that space can have for the military, and what it can do for them, what are the force multipliers.' He doesn't think there's much point in the spacemen trying to proselytize the rest of the military community about the importance of satellites: 'It's just too difficult, there are too few of us. Here I am at the National War College, and there isn't a lecture on the schedule associated with space. I'll bet you if I took a poll of the entire student body, there would be only a few at most who could tell me what space system provides tactical warning of a missile attack on the U.S.

'We've tried to educate, but the commands will not step up and say they need space support. For example, there was an argument once for conducting a Joint Chiefs of Staff [practice] operation without communications satellites. The JCS said, 'I can't do that, the operation won't be possible.' But yet they won't face up to the fact that that satellite is non-survivable. I think we should either make it survivable or get out of space. If we don't watch out, space could become our Achilles' heel without us even recognizing it.'

There's little question that satellite systems play important roles in the day-to-day peacetime operations of the U.S. military. The Soviets also rely on satellites, though, one might argue, to a lesser extent. If the United States were involved in a nonnuclear war anywhere on the globe, satellites would certainly provide a wide range of extremely useful, perhaps vital, services to our armed forces. The question is, how likely are we to become involved in a major war with the Soviet Union that does *not* very soon turn nuclear?

For example, U.S. officials have said that one of the first Soviet satellites we would go after would be their ocean reconnaissance spacecraft that could track U.S. aircraft carriers. What kind of war would we be in with the Soviets in which it was vital to them to destroy the backbone of our Navy, yet not vital enough that they should use nuclear weapons to try to do it? But if they were to use nuclear weapons, it's not at all certain that our carrier task forces would last very long, whether the Soviets used satellites to spot them or not.

In a Persian Gulf intervention against Soviet invasion, the 'force multiplying' effects of satellites would come in handy to our Rapid Deployment Force. A look at the map, however, suggests that the most efficient use of our troops could not compensate for the geographical advantage the Soviets would have in the area. It's generally recognized that the U.S. would have to resort to nuclear weapons at an early stage to hold back a Soviet force of any size. Our deterrent threat in the region relies not on Soviet fears of being beaten back by conventionally armed forces, but on their fears of the uncontrolled escalation of nuclear war.

A similar principle applies in Europe. There are varying opinions about how long NATO conventional forces would hold out against a Warsaw Pact onslaught, but the presence of 6,000 or so U.S. tactical nuclear weapons there says that if our side were losing, we would escalate. The planned deployment of Pershing and cruise missiles in Western Europe that could reach the Soviet Union are supposed to reinforce that threat. But if the Soviets, for whatever reason, ever decide that the time has come for war in Europe, why should they hold their own nuclear weapons in reserve until the West has had the chance to make the first strike? Or, in the most optimistic case of NATO conventional forces defeating the Warsaw Pact troops, why wouldn't the Soviets be as willing as we are to try to rescue the situation with nuclear escalation?

The next question is, can we have a 'limited' or 'theater' (that is, European or Southwest Asian) nuclear war? If so, the anti-satellite and satellite defense scenarios might still make some sense. But the Soviets, at least, deny this possibility. Part of *their* deterrent of *our* deterrent is that we would not be able to turn a war nuclear and escape the consequences ourselves. By planning to deploy those missiles in Europe that can reach Soviet territory, we are taking them up on the bluff if it is one. We are saying that *they* can't have a *limited* nuclear war either, that a nuclear attack on Europe will bring direct American retribution on the Soviet homeland.

In the case of general nuclear war, though – 'post-SIOP' as Heimach phrases it – do anti-satellite battles make much sense? Does it make any difference if we have our satellites or not? Those who think that 'protracted' nuclear wars are possible would say yes. They want to maintain that 'strategic connectivity' that keeps the nuclear weapons attacking long after the war has begun. They think that there will be a winner and a loser in the Big One, and that the winner will be the one who keeps his military forces functioning the longest.

If you take that possibility seriously, then the kind of preparations Heimach argues for might make some sense. On the other hand, there is not much point in building up our own anti-satellite forces *unless* we are planning to strike first, to start the nuclear war. The reason is that most or all of the earth-based men and equipment we would need to carry out anti-satellite weapon attacks will be lost to enemy nuclear missile attacks in the first fifteen minutes or half hour of the war.

The existence of nuclear arsenals casts a fog of uncertainty over all modern military planning. Military planners try to reduce this uncertainty by invoking 'kill probabilities,' 'radii of blast over-pressure,' 'equivalent megatons,' 'countermilitary potential,' and all the other conceptual paraphernalia of nuclear-age operations analysis. But the bottom line is that the sum of the effects of even 'limited' exchange of nuclear weapons on our military machine and our whole society is incalculable. And we're likely to have just one experiment to test out all the conflicting theories about what might really happen.

Despite the uncertainties in space warfare, some spacemen see space as an arena where we have a chance to outrace the Soviet Union, while others see it as an arena where we must compete just to stay even. But nearly all of them see the competition as inexorable. None of them takes seriously the possibility that the United States and the Soviet Union could come to a workable agreement to contain an arms race in space.

In 1978 U.S. and Soviet diplomats met to discuss the possibility of negotiating limitations on anti-satellite ASAT weapons. They held two more meetings, the last in Vienna in June 1979, when Carter and Brezhnev initialed the SALT II treaty. Another had been planned for February 1980, but in the meantime the Soviets invaded Afghanistan. The invasion ruined all hopes of Senate ratification of the SALT II treaty. It also led the Carter Administration to avoid the political risk of being seen negotiating with the Soviets during that election year. The ASAT talks went into limbo.

The Reagan Administration was even less willing to talk arms control with the Soviets. It had campaigned on a platform plank which asserted that the United States was militarily inferior to the Soviet Union, and that arms control agreements were the main cause of the inferiority. Without a U.S. anti-satellite weapon in being, with what was there to negotiate? And once the technologically superior Miniature Homing Intercept Vehicle had been tested and had given us the lead, why negotiate?

There is a military argument to be made for negotiating. United States military forces depend increasingly on space systems for valuable, sometimes vital, support. It might even be argued that, despite the large scale of the Soviet space program, the United States ultimately depends more on its space systems than the Soviet Union does. The Soviets keep most of their troops inside or near their own borders, so their lines of communication are largely internal. With troops flung across the Atlantic and Pacific, the United States depends much more on its communications, navigation, weather, and reconnaissance satellites. If the United States in fact depends more on its satellites, then it has more to lose in an anti-satellite battle.

The Soviet Union, true, has an ASAT that seems to work – at least part of the time. But it is a weapon with limitations. It's been tested only at limited altitudes and angles of inclination (and can't reach the high-altitude U.S. communications and warning satellites). It requires a big launching rocket and lacks flexibility. It would be difficult to deploy in large numbers. But in an unrestrained space-arms competition, surely we can expect the Soviet system to get better. The Soviet threat to U.S. satellites will grow.

On the other hand, if the two sides could agree to limit the competition, U.S. satellites might be *relatively* safer (the price would be the inability of the U.S. to do much against Soviet satellites). An agreement to stop testing and not to deploy anti-satellite weapons might not be 100 percent foolproof: it's conceivable the Soviets could find some subtle ways of cheating that avoided our methods of detection. It's also conceivable that the agreement could include cooperative measures to make it easier to monitor suspicious activities. For example, the two sides might agree to conduct orbital rendezvous of satellites according to certain 'rules of the road' that wouldn't be suitable for ASAT tests.

The essential question would not be whether we could guarantee that with a space-arms control agreement the Soviet Union could never destroy an American satellite. Instead, it would be whether U.S. satellites will be safer in an arms control environment or in an environment of an all-out anti-satellite arms race.

This argument suggests a possible *military* benefit from an ASAT arms control agreement. There are other potential benefits in which military men might be expected to take less interest. For example, keeping weapons out of space might close off yet another arena of armed confrontation where a war between the Superpowers might ignite. Or, at least, we might save the taxpayer from having to spend many billions of dollars in pursuit of a short-term military advantage, only to find that the battle of offensive measures and defensive countermeasures is endless.

Military men, however, tend to judge arms control measures only in terms of transient military advantage conferred on one side or the other. Major Lance Lord of the Pentagon Office of Net Assessments said: 'Whether you have an ASAT agreement or not, you're not going to be able to stop the confrontation in space, because there's a large momentum: technology is going to drag you there whether you want to go or not.'

That's the view of a military man who sees space as a place where we can get the jump on the Soviets, take a lead in the arms race, restore American superiority. Another spaceman, an expert in 'satellite survivability,' sees the issue a little differently. Colonel Heimach, a space warfare expert for twenty years, said he likes the idea of an agreement that would make U.S. satellites safer – he just doesn't think it's possible. 'Almost all of us agree that from the military standpoint it would be a super opportunity if we could get a treaty on space that was viable, because the support that comes from space, I think, is so great, particularly when you consider that it creates a force multiplier effect that helps us in our numbers disparity, that it would be something we'd welcome with open arms. But quite frankly, I haven't been able to figure out how to make a treaty that was viable where we couldn't get caught flatfooted. Therein lies the problem.'

I asked Heimach if, even with a treaty, the U.S. couldn't keep working on measures to protect its satellites against any Soviet weapon that slipped through the treaty monitoring system.

He said, 'If you came up with a treaty for banning an anti-satellite weapon, there is a concern that Congress would use the treaty as the justification for not financing those survivability efforts, and I think historically Congress has shown that once they got a treaty, they assumed that the other guy would be completely honest, and therefore we didn't have to spend any money. I think that chemical-biological warfare is a perfect example. We have an international treaty that bans development of those, and therefore we hardly went out and spent any money on protective clothing and all that other stuff, and in the meantime all the evidence points to the fact that the Soviets never adhered to it anyway. Most of us are afraid that the treaty would never result in the survivability measures because people assume that the treaty gives you survivability. Well, that's something that has to be argued out.'

This argument – that you can't trust Congress and the public to hedge against Soviet treaty violations – underlies much of the opposition to arms control agreements. In 1963, when President Kennedy sent the Limited Nuclear Test Ban Treaty to the Senate for ratification, he needed to get the Atomic Energy Commission and the Joint Chiefs of Staff to 'sign on.' They did so only in exchange for the promise of a vigorous *underground* testing program to make up for the aboveground tests prohibited by the treaty. In the first seven years after the limited test ban went into effect, the United States conducted twice as many nuclear tests as it had in the fourteen years before the treaty.

Military misgivings about the SALT II treaty ran along the same lines. It was hard to show that the SALT II agreements placed the United States at a military disadvantage relative to the Soviets, or that secret Soviet violations could somehow give them a decisive military edge. What many of the SALT II opponents really feared was that the American public and Congress would be 'lulled' into a false sense of security by the treaty, that they wouldn't put up the funds to modernize and expand U.S. nuclear forces within the limits allowed by the treaty.

Testifying before the Senators, the Chiefs had to admit that the SALT II agreements actually limited the Soviet Union more than it did the United States, and they said they supported the treaty *if* the country would build the new arms they wanted. The Reagan Administration, whose presidential campaigners had bitterly opposed the

SALT II treaty, ended up tacitly abiding by the agreements, hoping to keep the Soviets within the bounds established by the treaty.

So the real worry military men (and many of their sympathizers on the right) have is not about the Russians (whom they trust to do the worst), but about the public and Congress, who are too naïve. Colonel Heimach thinks it would be especially hard to get Congress to do what he thinks is necessary to hedge against a breakdown of an anti-satellite weapons treaty. He fears the possibility of a Soviet 'breakout' – a sudden repudiation of the treaty followed by the quick testing and deployment of new weapons.

He said that we were relatively safe agreeing to limit intercontinental ballistic missiles, because the Soviets would have to test them thoroughly and deploy a large number to get any advantage. With a crash program, we could match the new Soviet missiles and not be too far behind. But because the number of targets for ASAT weapons is so small – some space systems have only a few satellites – we'd have no time to respond. They could attack our satellites, but we couldn't get at theirs for three or four years.

Heimach used a ground-based anti-satellite laser as an example of something to worry about: 'There is no way that you could tell, through any intelligence means that I know of short of a man walking into the building, that a laser was under development, until he fired it the first time against a target, and even then you might not see it, but let's assume you could detect that. Once he kills a satellite, every other satellite within range of that laser is now hostage, because he can fire it again. And in general he doesn't have to make many tests to verify that that system works. For instance, the system that we had out at Johnston Island, we tested *three* times and declared it operational. . . . So my argument is that the treaty becomes almost useless – in fact may put you at a disadvantage because we *will* abide by the treaty – that's the other side of the coin. I approach it from that point of view, and historically we have.'

I asked Heimach if the Soviets wouldn't have to consider

the possibility that the United States too would prepare for a 'breakout.' In the case of the 1972 SALT I treaty limiting anti-ballistic missiles systems, for example, we have maintained an active research program on weapons whose deployment would be banned.

'But you see,' he said, 'I operate on the premise that we won't, because historically we haven't, and Congress hasn't, except in some extreme cases, funded things to get ready for a breakout. Now ABM is clearly an example, but there's testing going on, and it's allowed by the treaty.'

Heimach has clearly thought a lot about how the Soviet adversary could circumvent an ASAT treaty: 'He could develop a nuclear ASAT to go up to high altitude, and we would never be able to prove that he's done it, because you can conduct tests to points in space, and verify that you went through the point in space, and that's all you need. And all you have to do is monitor the closure of the switch before you detonate it.'

The United States, though, conducted such tests in the 1960s, and presumably we could reinstall an improved version of our Johnston Island nuclear ASAT system in the event of a Soviet breakout from a treaty.

Heimach has more ideas: 'Say he wanted to demonstrate that he could use the current system or an advanced system for other purposes: how do you prevent him from conducting a rendezvous on a satellite where he would say, "But for my national security I'm doing an inspection, testing an inspection." Therein lies the other side of the coin, because that is a valid argument for him: and I think eventually we're going to decide that we have to have an inspection capability also.'

But why not establish mutual 'rules of the road' regarding inspections – perhaps prohibiting them, perhaps setting limiting conditions on them, perhaps even conducting them with joint inspection teams? But then the U.S. military would be subject to the same kinds of limitations. And they're uncomfortable with limitations, even if the other side is subject to them.

Heimach again: 'Another thing that has to be treated separately is the issue of our ASAT as a deterrent to theirs.

Does the U.S. need an ASAT?'

But suppose the Soviets see us as more dependent on our satellites than they are on theirs? Will they really be deterred by our threats of counterattack?

'That's one of the arguments,' he said, 'but there are many situations where we may not be using space and where they may be totally dependent on it. If that situation comes up, you have to ask yourself if you want to do something about it. I think a lot of people tend to argue that you want to do something.'

Heimach has pondered more than most military spacemen on the question of space-arms control. But one wonders: if smart, thoughtful men like Heimach put as much effort into working out possible arms control measures as they put into figuring out why arms control won't work, maybe we'd get arms control proposals that *would* work.

But the fact is, precious little thought now goes into how we might draw the line on the militarization of space. From the spacemen, we get creative thought on how the military can more fully exploit space. The Air Force and the Pentagon hierarchy evaluate space programs according to their potential contribution to U.S. military power as opposed to their potential drain on resources needed for other military programs. Pentagon contractors – whether they are think tanks or hardware makers – propose what they hope their customers will buy.

The U.S. Arms Control and Disarmament Agency, now manned by ideological opponents of arms control, has been stripped of the handful of staff who took part in the Carter Administration's brief fling with anti-satellite weapons negotiations. Among those who speak out on military space policy on Capitol Hill, most urge the more rapid militarization of space; only a very few suggest the path of negotiated arms limitation.

It may in any case be too late to follow that other path. American spacemen are uncomfortable with the fact that the Soviets have already tested, repeatedly, an anti-satellite weapon (as, no doubt, Soviet military men were uncomfortable with the U.S. weapon based on Johnston Island until 1975). They think that the U.S. bargaining position in ASAT negotiations would be weak until the U.S. has tried out its Prototype Miniature Air-Launched System. But this system will be far superior to anything the Soviets have tested so far. Once we have it working, the *Soviets* will probably see themselves as being in the inferior bargaining position.

If the United States had really wanted to just catch up with the Soviets for bargaining purposes, it could have planned to test an orbiting 'killer satellite' more like the Soviet weapon. In fact, the Air Force is quietly working on just such a system as a backup to the technically more ambitious Miniature Vehicle. But in this case the United States isn't arming to parley: it's arming to get an edge.

If we're going to treat space as first and foremost a new arena of military competition, let's have no illusions about it. Space technology is not a *deus ex machina* that will resolve our frustrating strategic stalemate with the Soviet Union. It's true that satellites can help our land, sea, and air forces in ways unprecedented in past warfare: providing instant global communications, timely weather information, comprehensive reconnaissance, ultra-precise navigational data.

But it's also true that the more military capabilities we put into satellites – the more of an edge they seem to give us, the more dependent on them we become – the more those satellites themselves become inviting targets to Soviet military men.

If they think we have something that will let us beat them, they'll go after it. If they want a satellite badly enough, they'll probably be able to get it. We can't protect it, we might be able to hide it, we might be able to replace it, we might be able to raise the cost of getting it (by, say, putting up many redundant satellites). But 'denial' is always going to be easier than 'maintaining free access'; if super-sophisticated technology doesn't make that so, nuclear weapons do.

# CHAPTER EIGHT

# Space Lasers: The Ultimate Weapon?

The most optimistic spacemen believe that the U.S. can defend itself from nuclear attack by building a fleet of laser-armed satellites. Some argue that one laser battle station could protect U.S. military satellites within a radius of 3,000 miles. Four stations could keep all, or nearly all, Soviet bombers, as well as some submarine-launched ballistic missiles, out of the United States. Twenty-four stations might frustrate a Soviet counterforce strike by knocking out their big SS-18 ICBMs. More ambitious schemes call for a hundred or more battle stations in space. Proponents argue that laser weapons might end the nuclear balance of terror by sealing off the United States from Soviet nuclear attack.

The most ambitious schemes of all project the use of high-energy, subatomic particle beams in the anti-ballistic missile role. Particle-beam weapons in space would require giant nuclear accelerators of unprecedented power. While the Pentagon is devoting some money to research on such weapons, no credible expert seriously argues that anything resembling a particle-beam battle station could be operating in space in the foreseeable future. Building laser weapons would be a formidable enough task for this century.

The Defense Department is betting about $150 million a year that space laser weapons are technologically feasible. There are three major lines of research that will have to be successful if the Pentagon is to get a usable weapon. First is the laser generating device itself; second is the mirror for focusing the laser beam on a distant target; third is a system for finding, tracking, and aiming the beam at the targets.

There are several different techniques for producing the ultra-concentrated beam of light known as a laser. These techniques produce light at different wavelengths (the shorter wavelengths, being more energetic, are preferable) and with varying efficiencies in translating the energy source into lased light. The Defense Advanced Research Projects Agency, DARPA, has a program called Alpha designed to produce a 5-megawatt (5-million-watt) laser for a space platform. It could damage other satellites and perhaps high-altitude aircraft. Estimates vary, but a full-scale anti-ballistic missile laser with a range of better than 600 miles might require a 25-megawatt laser 'pump.'

Kosta Tsipis is an MIT physicist who has taken on as his special mission the task of bridging the gap between the military technicians on the one hand and the 'peacenik,' or pro-arms-control, community on the other. He heads the MIT 'Program in Science and Technology for International Security,' which over the years has issued detailed technical studies of such questions as the 'hardness' of Minuteman ICBM silos and the feasibility of space laser weapons.

Tsipis has calculated that a 600-mile-range space laser would have to put out 100 megawatts of energy to destroy an aluminum-skinned missile in about one second. If we assume optimistically that the laser device could convert 30 percent of its fuel's energy into laser energy, then it would take about 4,400 pounds of fuel to destroy one missile. But, depending on its range, each laser battle station would have to be able to shoot down several hundred, perhaps a thousand missiles in a few minutes. A fleet of fifty such battle stations would then require more than 100 million pounds of fuel. It would take 2,750 space shuttle trips to deliver that much fuel into polar orbits. Laser enthusiasts generally agree that we'd need a new space transportation system, capable of much heavier loads than the space shuttle, to put up a space laser ABM system.

The only time Soviet land-based and sea-based ballistic missiles would be vulnerable to space laser attack would be during the period in which they have cleared most of the atmosphere and in which their rockets are still firing – the boost phase. This phase lasts about five to eight minutes. If the laser can stay focused on the missile long enough (how long depends on how powerful and well aimed the beam is), it can melt the skin and probably throw the missile off course.

The Army, which runs most of U.S. ABM research, has been carrying out experiments both on laser sources and on what happens to the targets of lasers. Late in 1981, for example, Lockheed and TRW each got Army contracts to study the vulnerability of missiles to lasers. Early in 1980 Richard Carlson, director of the Army's Ballistic Missile Defense Advanced Technology Center, pointed out that just what would happen to ICBM's in a space laser attack was still 'an unknown that has to be well understood.'

In research for the Navy, the Vought Corporation has already come up with an ingenious technique that might protect missiles and aircraft from laser damage. A highly polished aluminum surface would reflect 97 percent of the energy from an infrared laser beam. But exposed aluminum oxidizes in air, so a high polish won't keep. Painting the aluminum is even worse, since paint carbonizes under the laser heat, blackening the metal surface and making it even readier to absorb the damaging laser energy. The Vought solution is to coat the aluminum with what is called an 'ablative' substance, a material that evaporates away under the laser beam, leaving the surface of metal still highly reflective.

Tsipis has suggested other laser countermeasures: coat the missile with some highly reflective material; have the missile secrete a layer of fluid from its nose, absorbing the laser energy in the way that an ablative coating would; simply spin the missile, reducing the amount of time that any one section of its surface is exposed to the laser beam. Whether or not any one of these counter measures turns out to be wholly effective, it makes sense to investigate and understand them before we commit many billions of dollars to the weapons.

Generating enough laser power to knock out missiles is just part of the task of putting together a space laser ABM system. To focus and aim the laser beam at targets from 600 to 3,000 miles away (depending on the particular features of the system) will require enormous mirrors of exquisite optical quality, extraordinary durability, and ultra-light weight. For anti-satellite and anti-aircraft purposes, the mirror would probably have to be at least 4 meters (13.1 feet) in diameter; an anti-ballistic missile weapon would probably need at least a 10-meter mirror. A DARPA draft report, leaked to *Aviation Week* early in 1981, estimated that the kind of complete coverage against ballistic missiles that the most extreme laser enthusiasts envision would require a system of one hundred satellites, each with a 25-megawatt laser and a 15-meter (49.2-foot) mirror.

Tsipis wrote, 'Making such a mirror sufficiently rugged and of the necessary optical quality . . . is beyond the technical capabilities of the U.S. or any other nation. There are scant prospects for constructing an optically precise four-meter mirror.' Senator Malcolm Wallop of Wyoming has attempted annually to get more money devoted to space lasers and to force the Pentagon to move faster toward putting a demonstration weapon into orbit. He differs with Tsipis, asserting that 'there has never been any doubt that such a mirror could be built . . . The research has paid off to the point that a major aerospace company has proposed a fixed-price contract for delivery of a segmented ten-meter laser mirror in five years.' The company is United Technologies, which wants $80 million for the mirror. What Wallop *didn't* mention was that United Technologies admits there are several 'unresolved issues' in the technology to go into such an unprecedentedly large mirror, and that the mirror they propose is a 'demonstration' model. Before they build that one, they propose to build a smaller one 'to resolve any remaining basic technology issues and the unique fabrication and polishing' that such mirrors would require.

Supposing the weapons designers manage to scale up a laser source to unprecedented power and a mirror to unprecedented size and precision, the weapon still has to find the target. Shooting at a missile from 3,000 miles in space is like aiming from New York at a garbage can over Los Angeles. In the field of view of the laser battle station, the target will be about one one-hundred-thousandth of a degree (or less than one millionth of a radian) across. The weapon's sensors have to locate that little speck, then the laser-beam-forming mirror has to slue to track it, perhaps for several seconds. (Each battle station would have to do this for several hundred, perhaps a thousand missiles, within a few minutes, depending on the range of the weapon and the total number of stations.) Fortunately, the target is a flaming rocket, spewing out hot gases detectable by infrared telescopes and sensors.

In 1981, a highly placed Defense research official told the Senate Armed Services Committee: 'Advanced space-based surveillance systems, not currently planned, would be necessary to detect and track large numbers of targets for significant space-based laser missions. The separate surveillance systems required for aircraft or ballistic missile targets would be major systems developments in their own right. While the technology for such surveillance is under development, there are no current plans to deploy such systems.'

The target detectors for the space-based laser ABM system will have to do more than pick out rocket flames at anywhere from 600 to 3,000 miles away. The Soviets could spoof the system by firing off thousands of flares just before launching their rockets. It might be possible to design sensors that could sort out the real missiles, perhaps not by merely detecting the sources of infrared radiation, but by the more difficult optical operation of resolving a focused image of the missile. Then, however, the Soviets have the option of deploying hundreds of dummy missiles, cheap to make because they'd lack the costly nuclear warheads and precision guidance systems of the real missiles. Another option might be to post ground-based lasers in the missile fields. They wouldn't have to be too powerful: they would just try to blind the laser battle stations' sensors.

Then there are the 'C-cubed' – the communications, command, and control – problems. The perfected target detector on each battle station has to turn the targets over to a tracking system linked to the mirror directing the laser beam. A small companion tracking laser would probably have to do the job. Probably in 1987, the Air Force and DARPA will send up on the space shuttle an experimental laser pointing and tracking device as part of a project code-named Talon Gold. The Talon Gold device, to be built by Lockheed, will try to point a low-power laser at airplanes and satellites, not missiles. The experimenters hope to track the targets to within two tenths of a millionth of a radian (a radian is a little over 57 degrees).

Even if the tracker is good at aiming the laser beam at a rocket, it can't be perfect every time. There has to be a mechanism for calculating miss distances and reaiming the laser. There's the additional problem of preventing 'beam jitter,' which is just enough fluctuation of the beam off a single spot on the missile to spread out the energy deposited there, prolonging the time necessary to damage it fatally. And then there should be some system for determining whether the missile has been damaged enough to send it off course – unless there's absolute confidence that a hit is as good as a kill.

The problem gets more complicated when you consider that a complete system would need anywhere from twenty-five to several hundred battle stations. Somehow, those stations in range of the same sets of missiles will have to decide, automatically, how to divide up the targets among themselves, as well as pass targets along as each station moves in and out of range. Relative to the surface of the earth from which the missiles are launched, each laser station at an altitude of 3,000 miles would be moving along at about 13,300 miles an hour. Stations at 600 miles up would be moving at 16,500 miles an hour. Any one battle station's target area would be shifting constantly, with a complete turnover from one statellite to the next every fifteen minutes. A system designed to cover continuously all the areas in the world from which the Soviet Union could launch missiles would need more than 400 satellites if the laser range was 600 miles; it would need more than 25 if the range was 3,000 miles.

The 'command and control' of these battle stations would be so complex, and would have to be done so quickly, that human intervention would be impossible. Speaking to the Senate Armed Services Committee in 1981, Pentagon laser weapons research expert George Millburn said, 'Another risk is operational: how would we operate such a system . . . It is a global system. In many cases the command and control would have to be autonomous. We would have to delegate the decision-making to the weapon system itself and we have no experience in that type of operational system.' Developing that automated control system would in itself be a major project. And if we can do it, do we then let the system automatically shoot down any nation's rocket launches? Do we try to require all other countries to notify us in advance when they plan to launch a rocket so that we can turn the system off for them?

Assuming for the moment that the space laser turns out to be technologically feasible, will it be economically feasible? It's too soon to make firm estimates about the ultimate costs of the space battle fleet, but the Defense Department has come up with some rough figures. A system designed to attack other satellites and high-altitude bombers could cost about $50 billion. A system designed to offer limited protection against some Soviet ballistic missiles – say, to dilute an attack on U.S. missile silos – might cost about $100 billion. A full-scale anti-ballistic missile system, designed to offer the kind of protection against all Soviet missiles that space laser enthusiasts endorse, would cost about $500 billion.

And that wouldn't be all. According to George Millburn, it is very unlikely that a space-based laser system by itself could ever seal out Soviet warheads. At the 'leakage rate' that is likely (his estimate is classified), he said, '. . . if we considered the Soviet numbers of re-entry vehicles they could fire at us, the numbers that could penetrate through that defense could cause significant damage to the United States. It is unlikely we would consider doing a space-based laser ABM system without having an overlay system and most particularly . . . an adequate civil defense capability to back it up.' So the $500 billion, even without the inevitable aerospace industry cost overruns, would be just for starters.

Verne Lynn, Director of Defensive Systems for the Under Secretary of Defense for Research, took part in a Defense Department study of the potential of space lasers. He told the Senate Armed Services Committee, 'There is no question but what the compelling motivation would be to build a system that would get us out from under the nuclear 'high noon' that we have now. . . . That is why the damage-limiting BMD [Ballistic Missile Defense] is attractive. The thing that concerns me is that there are enormous uncertainties.' One of the principal uncertainties, he said, is the 'survivability of the laser.' If the system were so valuable to us, it would be a correspondingly tempting target to the Soviets. Lynn explained, 'The laser in this case could threaten the strategic balance. It would be an extremely high-value target. Again, there are no returns, but it is easier to develop threats to the laser than it is to solve them.'

Millburn later said, 'The survivability of the laser weapon system against a concerted enemy attack is a matter of concern. There is little doubt that, like most weapons systems, space-based laser battle stations could be destroyed by a dedicated attack.' He added that the 'complex interaction' of offensive and defensive measures 'has not been examined.' Milburn cited several potential threats to the laser battle stations. First was the current Soviet anti-satellite system. His testimony here was censored from the public record, but because the Soviet system relies on a large ground-based rocket, it is probably not too great a threat. The trick would be to identify the purpose of the Soviet launch before the Soviet killer satellite got too close.

Millburn talked about several other possible threats to a space-based laser system. For example, the Soviets might build a mirror image of the U.S. Miniature Homing Intercept Vehicle anti-satellite weapon. That would pose a much greater threat, even against an operating space laser, than the current Soviet system. The Miniature Vehicle can be launched by a fighter aircraft or lofted with small boosters that would be much more difficult to detect than an ICBM or SLMBM launch. He pointed out that the battle station would also have a hard time finding the small vehicle itself, and that the little interceptor could approach from almost any direction, including one from which the space-based laser's sensors are blind. The Miniature

Vehicle could also be launched in salvos, making it hard for the laser to counterattack even if it could spot some or all of them.

'Another threat to the space-launched laser is the "space mine" or "fellow traveler." This concept is a conventional or nuclear weapon launched into orbit accompanying the prospective target.' The Soviets could either explode the mine on command or instruct it to explode automatically. Said Millburn, 'The space-based laser system's problem then would be to prevent the mine from ever getting within lethal range in the first place. Assuming that these mining tactics can be recognized when first employed near individual space-based laser stations, the U.S. would have to enforce a sterile zone around each space-based laser battle station.' Millburn made another assumption: that the Soviets do not plant their 'mines' before the laser station ever becomes operational and able to defend itself, rather than waiting until it has been fueled up and checked out.

He also mentioned the possibility of a Soviet space-based laser that could attack and outfight a U.S. one, but he said that 'perhaps the most serious threat to the space-based laser battle station is a direct ascent, one-on-one, nuclear anti-satellite, presumably heavily protected with an ablating heat shield to minimize the effective keep-out range of a self-defending space laser.' What we don't know here, Millburn said, is whether you could armor the battle station well enough to protect it from the intense X rays of a nuclear explosion in space.

There are some proponents of space laser anti-ballistic missile (ABM) systems among the uniformed military spacemen, but the most vehement advocates have been in Congress and the aerospace industry. In the Senate, there is Malcolm Wallop of Wyoming. In the House, Representative Ken Kramer has become an active laser advocate. Behind the scenes, the Congressional advocates and their staffs have received advice from employees of companies involved in laser research – companies like Rockwell, Boeing, TRW, Bell Aerospace Textron, Hughes, United Technologies, Lockheed.

Since the mid-1960s U.S. nuclear strategy has been in the

'iron grip' (Malcolm Wallop's words) of a foolish and immoral 'doctrine' known as Mutual Assured Destruction, or, appropriately, MAD. This doctrine supposedly requires that our only defense against a Soviet attack be the threat to retaliate with the mass nuclear destruction of the Soviet population. According to MAD, nuclear weapons are so terrible that no rational military use can be made of them, and therefore our only protection is a balance of terror.

A corollary of MAD is that active defenses – for example, anti-ballistic missile systems – and passive, or civil, defense are unworkable and only appear provocative to the other side. Thus the ABM treaty of the SALT I agreements, signed in 1972, attempts to assure that the ability to kill many people – 'assured destruction' – will remain mutual.

Strong proponents of space-based weapons deem the doctrine of MAD as immoral. It makes our primary strategic objective the mass destruction of people rather than the winning of war. It keeps the United States from defending the people of the country from the ravages of nuclear war if deterrence breaks down.

Two Air Force advocates of 'initiating wide-open military competition in space as a way of moving beyond present deterrent strategies,' Lieutenant Colonel Barry Watts and Major Lance Lord, are unhappy with the 'corrosive effects that living with the balance of terror have increasingly had on our national spirit.' They say, '. . . for American decision-makers, the very unthinkability of even limited nuclear use has, over time, subtly affected our capacity to contemplate nonnuclear conflict as well. Because the threshold beyond which combat involving U.S. or Soviet forces could irretrievably escalate to uncontrolled nuclear exchanges remains highly uncertain, it is but a short step to the thought that any war directly involving superpower interests could become a nuclear war. The psychology of the balance of terror, therefore, has made it more and more difficult for democracies like the U.S. to accept the risks of employing force at any level. The longer we have lived with deterrence through assured destruction, the more our preoccupation with preventing war has eroded our will to defend ourselves or our interests.' Putting laser anti-ballistic missile weapons in space, they conclude, would let us 'leave the missile age behind' and 'divest ourselves of this creeping paralysis of will.'

To this school the MAD doctrine rests on the incorrect assumption that there is no defense against ballistic missiles. The laser advocates say that if superior American technology is applied to the problem, we *can* defend ourselves. On the other hand, if we don't, the Russians will. In a 1979 article Wallop argued that space-based laser weapons ". . . at least hold the promise of barring nuclear-tipped ballistic missiles of mass destruction from the arena of war. To be sure, the superpower that grasps that promise first will wrest an enormous strategic advantage.'

Laser advocates also think that Pentagon experts are simply committed to the 'doctrine' of Mutual Assured Destruction, and therefore are against any ABM system on blind principle.

One of the targets of that kind of criticism is Dr Richard Airey, who has been the Pentagon's Director of Directed Energy technology under both the Carter and Reagan administrations. Asked about MAD, Airey conceded that until recently the technological obstacles to effective defense against ballistic missiles had led U.S. military planners to emphasize offensive weapons, though he thought that with the arrival of the Reagan Administration the subject of ballistic missile defense was getting a new look.

As to whether space laser weapons offer us a way out of the nuclear balance of terror, Airey said simply, 'I don't think anybody knows yet. . . . We don't understand the technology well enough, we don't understand the system concept well enough in order to be able to say for sure.'

Airey explained that many chiefs of the aerospace industry, who are responsible to the Defense Department concerning the laser program, missile development, and all other areas, tend to conclude, as does the Secretary of Defense, that 'we can't just put all our eggs in one basket when there's so much uncertainty involved.'

In July 1981, Angelo Codevilla, Wallop's staff adviser on

lasers, spoke to a conference held by the American Institute of Aeronautics and Astronautics. He said that with $200 billion the industry could build a space laser system in ten or twelve years. He complained that if only the industry would show enough courage to declare its willingness to take a chance on space lasers, Congress and the executive branch would fall in behind.

Richard Airey retorted, '. . . I totally agree with Senator Wallop's view that *if* there's a possibility of building a weapons system which can defend the United States and its population, that we ought to try hard to build such a thing. But I think that he's not a technologist who's been steeped in this particular technology for many, many years. It's a *chance*, not a certainty. So for people who don't know the technology well to accuse those in the business of dragging their feet, of not caring, and so forth, is quite incorrect. We care very much about this, and within the limits of competing budget process we're trying as hard as we can to find out the answers so that we can say, sometime in the future, "Yes, it really does make sense" or "No, it doesn't make sense." '

Mutual Assured Destruction as a 'strategy' or even a 'doctrine' is a straw man. The myth that MAD has been the central principle of U.S. nuclear strategy is traceable to the unrelenting hostility felt by military hard-liners, in and out of uniform, toward Robert McNamara. McNamara, Secretary of Defense under Presidents Kennedy and Johnson, and his 'whiz kids' (of whom Carter Defense Secretary Harold Brown had been one) have never been forgiven for their attempts to impose managerial rationality on the previously intuitive art of military decision-making.

McNamara and his systems analysts did identify, in their words, 'assured destruction' as a *condition* but not necessarily as a *strategy*. In McNamara's Pentagon, 'assured destruction' meant the undeniable ability to inflict, *at a minimum*, an unacceptable amount of nuclear damage on the enemy's people and industry. McNamara thought that the minimum levels of destruction should be about 25 percent of the population and 50 percent of the industry. The prospect of such devastation should be enough to deter any rational Soviet leader from launching an attack on the United States. Therefore, at least this much destruction should always be *assured* even if the Soviets struck first. If the Soviets were to improve their civil defense, or to build up defensive weapons systems, then it would cost us more to maintain our guarantee of 'assured destruction' against them, since we would have to find ways to neutralize those systems. But we would do it.

By the same token, there was no practical way to deny the Soviets a similar assured destruction capability against the United States. Imperfect missile defense systems would just stimulate a Soviet offensive buildup to maintain assured destruction. Assured destruction, as long as both sides were determined to maintain it, was bound to be mutual. Assured destruction was and is not a 'doctrine': it is a fact of life.

It was in acknowledgment of this fact of life that the United States and the Soviet Union negotiated the ABM treaty of 1972. The point was not to guarantee to the Soviets that they would always be able to destroy the United States: the Soviets would take care of that themselves, with or without the ABM treaty. The point was to limit the stimulus to piling up offensive arms that ABM systems threatened to bring.

SALT I was a disappointment, though, because neither side ever limited its own offensive forces to the minimal requirements of maintaining an assured destruction capability. Even as McNamara explained the idea of assured destruction, the U.S. possessed several times as many nuclear warheads as would be needed to carry out that mission, even under the worst of assumptions. And things went up from there. The development of multiple independently targetable warheads, MIRVs, was supposedly a hedge against Soviet ABMs. But even after the ABM treaty, U.S. 'MIRV-ing' proceeded apace. The Soviets followed suit.

Any Defense official familiar with the SIOP (the U.S. nuclear war plan) will tell you that the military have never limited themselves to 'targeting' only cities, people, or industries. Rather, missile silos, airfields, barracks, military transportation nodes, weapons storage areas, and a hose of other military targets have long been in the SIOP. In other words, 'assured destruction,' while stating the minimum requirements for U.S. nuclear forces, has never been the *strategy* guiding war plans or weapons buys.

ABM supporters find it difficult to accept the *condition* of mutual assured destruction. Acknowledgment of this condition makes it difficult to define what a victory in nuclear war would be. There is despair over the possibility of victory. The hard-liners drag out numerous quotations from Soviet military officers seemingly demonstrating that they believe in victory. They tend to ignore all the other statements by high Soviet officials indicating the futility and universally disastrous consequences of nuclear war.

The pro-ABM school believes that American technology can rescue us from the mutuality of assured destruction. For them, the space-based laser ABM promises to take the assuredness out of the Soviet threat to destroy us so long as we get those weapons first. Proponents think we could stop them by putting up our own and shooting down theirs. Other students of the problem think there's no point in even trying.

Though not an Armed Services Committee member, Senator Wallop was allowed to sit in on the laser hearings. James Wade, a Pentagon laser weapons research expert (and George Millburn's boss), disagreed with Wallop's claim that 'technically, we are really not talking any major breakthrough necessary to get ourselves a fully functional, capable space-based laser, that there are engineering problems, but there are not frontiers there to be breached. . . .'

Wade replied that the 'question of how to handle a nuclear weapon employed in space against a laser battle station is a serious and yet unanswered question,' as is the issue of using the laser itself to defend the station. Wade continued, 'A nuclear weapon has the most efficient energy density of any weapon you can employ in space right now. I am not saying we can't defend against a nuclear weapon in space somewhere downstream, but we cannot do it now. We should be doing more research here.'

Space laser advocates often raise the Soviet specter: if we

don't rush ahead with the weapon, the Soviets will get there first. With laser battle stations in place, they will be able to seal out our nuclear weapons. Our assured destruction capability against them will be lost.

In 1982 Representative Ken Kramer (Republican from Colorado Springs) inadvertently leaked an estimate by the Pentagon research chief, Richard DeLauer, that the Soviets might have a laser weapon in space by the mid-1980s. A few days later, Air Force Chief of Staff Lew Allen, who ought to have been awfully concerned about such a development, said he didn't believe it. Even De-Lauer himself wasn't terribly alarmed by his own warning. A year earlier, the Reagan Administration fiscal 1982 budget for space laser weapons research was set at about $150 million. Not only did DeLauer fail to ask for a supplement to that amount, but the 1983 funding request was nearly the same.

At the end of 1979, Richard Airey had said that the Defense Department did 'not regard itself to be in a race for laser weapons.' He said that the pace of the U.S. program had been '. . . dictated by the advance of the technology and budget priorities, and not by direct competition with the Soviet program.' He acknowledged that the Soviets were spending a lot of money on lasers, and might even be trying to build weapons. In 1981 the consensus of Pentagon officials testifying on lasers before the Senate Armed Services Committee was that if the Soviets were to try to put a laser up in the next few years, they might score some propaganda points, but they were quite unlikely to have a workable weapon.

A limited U.S. space laser, deployed before its time, might or might not be able to challenge a similar Soviet weapon. What could challenge that weapon is the new U.S. anti-satellite system, the Miniature Air-Launched System. The Pentagon analysts, after all, don't know how they'd defend a space laser of our own against such a weapon.

It's also important to note that all of the other kinds of counter-measures described above that would be available to the Soviet Union would also be available to the United States. We wouldn't need a space laser of our own just because the Soviets had built one.

Disappointed with the reluctance of the Pentagon to lay out more money, space laser boosters have reacted to Defense Department reasoning on lasers by questioning the motives of the bad-news bearers. *Aviation Week*'s Clarence Robinson quoted anonymous 'Senators and staff attending the [1981] hearing' as labeling the Defense officials '. . . holdovers from the Carter Administration who have displayed hostility toward laser weapons in the past.'

Robinson ran a one-man campaign in the pages of his magazine for intensified space laser efforts. His reporting on this issue makes for an instructive case study in the selective revelation of supposedly classified information leaked by Pentagon, industry, and Capitol Hill officials with policy axes to grind. In February 1981, Robinson opened an article with 'Early technology demonstration of space-based, high-energy laser battle station through an accelerated program is urged in a Defense Dept. report soon to go to Congress. Laser battle stations in space offer the potential to alter the world balance of power, according to the report.'

Robinson quoted freely from this classified report, which has never been released to the public. But nowhere in his article did he mention the fact that the report was never meant to be an official Defense Department report, but rather a draft report only from the Defense Advanced Research Projects Agency (only three months later, in another article, did he mention the agency authorship of the document). Pentagon directed energy research chief Richard Airey later pointed out that the quoted document was only the internal contribution by DARPA to the actual Defense Department report that later was given to Congress.

Despite the subtitle, 'Defense Dept. report urges early effort on space-based battle stations,' the Robinson article's paraphrasings and quotations from the report suggest otherwise. He wrote that the report 'lists the estimated earliest times for space-based laser systems of given capabilities.' A single test weapon would take nine years to get into space. A ten-weapon system that could attack satellites, high-altitude aircraft, and some ballistic missiles would take fifteen years. A constellation of one hundred large stations for full-scale ballistic missile defense would take twenty to twenty-five years.

A system half that size might cost $140 billion, but, as Robinson paraphrased the report, 'There are still uncertainties in the data base for costs, times, performance and vulnerabilities, which levy urgent requirements on the technology program to resolve the issues of both weapon performance and countermeasures effectiveness.' The report also pointed out the need for new surveillance systems and a space-launching capability far exceeding that of the space shuttle.

After first intimating that the 'Defense Department' report supported his position, Robinson went on to quote unnamed 'laser experts and members of Congress' who objected to the pessimistic assumptions made in the report. For example, one of these said the document called for '. . . a perfect laser system, just as we have done in recent years seeking the perfect tank or the perfect bomber while the USSR went ahead and fielded lots of less-than-perfect tanks and bombers.'

In March 1981, Robinson reported on the Senate Armed Services Committee hearing where the top Pentagon laser officials testified on the inadvisability of a crash laser program (the one at which they estimated that a full anti-ballistic missile system might cost $500 billion). Despite the actual conservatism of the DARPA report as he himself had summarized it (and despite the fact that the record of the hearing itself was at that time still classified), Robinson reported the Defense officials' testimony as being 'in direct conflict' with the DARPA draft report, which he now described as 'clearly stating that the technology is available to alter the balance of power with the Soviets.'

In May 1981, the Senate again rebuffed a Wallop effort to force an extra $250 million on the Pentagon for the space weapons, but it did order an extra $50 million to be shifted to space lasers from other Defense budget items. (The list of Senators supporting the $50 million amendment read like a Who's Who of the Senate right wing: Warner, Goldwater, Symms, East, Helms, Hatch, Armstrong, Dole, Laxalt, Simpson, Inouye, Hayakawa, and Tower; Harrison Schmitt also spoke in favor of it.)

Apparently some DARPA officials had been 'scheduled' to brief the Reagan National Security Council on space lasers, 'in time for the White House to add its approval to the $250 million amendment and to have the Defense Dept. program offsets identified' before the Senate vote. Considering that the Reagan Administration was no more willing to request a total of $350 million on space lasers the next year than it was that year, the optimism about White House approval was obviously unfounded. In any case, the Defense Department did not approve the BSC briefing in time. 'One Senator' complained to the *Aviation Week* reporter that '. . . a Pentagon official from the Carter Administration held over in the Defense Dept. is opposed to laser weapons and held up approval of the council briefing.'

Robinson also reported that a Defense Science Board panel had failed to submit its study on lasers to the Senate Armed Services Committee in time for markup of the budget authorization bill. But it had, he wrote, briefed some Senators a day before the Wallop amendment went to the floor. Another example of *Aviation Week*'s selective reporting on the subject is that it made no mention of the fact that the Defence Science Board panel basically supported the Pentagon position on space lasers. The head of the panel, former Pentagon research director John Foster, concluded: 'In our view, it is too soon to attempt to accelerate SBL [Space-Based Laser] development toward integrated space demonstration for any mission, particularly for ballistic missile defense.'

Senator Warner, though a supporter of the Wallop amendment, included in the record his own views on the laser weapons. His memo read: 'In the 21st Century directed energy weapons such as space lasers are almost inevitable, but achievement of an effective space-based ballistic missile defense system is far more difficult and expensive than the most extreme enthusiasts admit.' He pointed out that numerous difficulties needed to be resolved in almost every aspect of laser ABMs. The Science Board panel had recommended that an average of $50 million a year be added to the space laser program.

Warner was particularly skeptical of the notion that lasers were going to free us from the nuclear balance of terror: 'Some advocates have suggested that space lasers will make offensive weapons obsolete, thus providing a new arms control regime. In fact offensive and defensive weapons always work together and in this case adversaries, unwilling to live without an offensive capability, would undoubtedly plan to attack space lasers with ASAT systems including other space lasers so as to free their offensive forces.'

What Warner didn't point out is that even if, by some remote chance, space lasers should fulfill their supporters' promises and put up an effective defense against ballistic missiles, that defense would hardly put nuclear weapons out of the picture. Neither side is about to let its 'assured destruction' capability go up in smoke as long as the other side has it. The Soviets might be forced to shift their strategic designs from 'nuclear war-fighting' – so-called 'counterforce' weapons – but if they did, it would be a shift *toward* maintaining the mutuality of assured destruction.

Even a defensive system that was highly effective against ballistic missiles would not be 100 percent leakproof. Some small number would certainly get through. But the Soviet target planners wouldn't know *which* of their warheads might get through, so they would have to hedge their bets. Large numbers of 'military' targets would now be out. Cities would still be in. The Soviet planners might not know which cities they would succeed in hitting, but they could assume that one is roughly as valuable to the United States as the next. The point is to make the prospective damage unacceptable to us so as to deter us from risking nuclear war with them. That is the bottom-line reason for having nuclear weapons.

Nor would the Soviets be limited to a drastically weakened ballistic missile force. Ballistic missiles and high-altitude bombers are not the only ways to deliver nuclear weapons. There is, for example, the cruise missile. Cruise missiles may be slow, but they fly low, and space lasers couldn't damage them even if they could spot them. If the high-flying airplanes that launch cruise missiles can't be designed to fly below laser range, the cruise missiles themselves could be built to the necessary fuel capacity to go from one continent to another. Otherwise, the Soviets could put hundreds or even thousands of cruise missiles on submarines (the U.S. already plans to do this). They would still be able to assure our destruction, and we theirs.

Occasionally the anti-ballistic missile system supporters back off from their rhetoric about the immorality of MAD. Conceding that the Soviets might be able to build laser-resistant missiles, they argue that it would at least be worthwhile to force the Soviets to spend a great deal of money rebuilding their rocket force.

At the March 1981 Senate hearings, Wallop tried this argument on DARPA directed energy chief Douglas Tanimoto, who thoroughly agreed with Wallop.

But at this point a 'holdover from the Carter Administration,' Seymour Zeiberg, spoke upo. Zeiberg, who had stayed on temporarily as Pentagon research head for strategic and space systems, said that Pentagon analysts had considered the trade-off between the cost to the Soviets of 'hardening' their missiles on the one hand and the cost to us of building the lasers on the other. Although Zeiberg's testimony was partially censored in the record, it's clear that the trade-off was unfavorable. The Soviets could protect their missiles with 'a relatively straightforward retrofit,' and they could do it faster than we could put up the kind of laser system that would be feasible in the next few years.

So the early deployment of space lasers would *not* force the Soviets to scrap their current missiles, or even cost them a great deal to add some protective coatings. But even if it did, forcing the Soviets to spend more money on new missiles is one thing; extricating America from the nuclear balance of terror is quite something else. Lasers notwithstanding, there is still no technological fix that is going to rescue us from the peril we share with the Soviet Union – the assuredness of our mutual destruction in a nuclear war.

Until 1983, the Reagan Administration had not decided to devote much of its enlarged military budget to a crash space laser program, but it had increased the annual research expenditure from about $100 million a year to $150 million. The Wallop amendment to the fiscal 1982 budget (that is, the one passed in May 1981) not only added $50 million to space laser research, but also directed the Air Force to set up a 'program office' to at least come up with a design for a complete space laser battle station.

On March 23, 1983, President Reagan gave a television presentation soon dubbed the 'Star Wars Speech' by the press. In the speech, he called for a new research program to seek technology capable of neutralizing nuclear weapons by sealing out enemy ballistic missiles. Many took this to be a call for moving toward the space laser ABM. A special committee to advise the President on defensive technologies was formed under the direction of former NASA chief James Fletcher. It began to look as though the space laser advocates were coming into their own – although the amounts of money to be devoted to their cause still remain uncertain.

The debate will continue in Congress over how much money to spend, what kinds of research to sponsor, and whether to insist on putting an early 'demonstration model' into space. The war on the ground in Washington over the future of laser battles in space has just begun.

# EPILOGUE

The military exploitation of space has been going on for twenty-five years. With most decisions still secretly left to experts, and an aura of esoteric complexity surrounding the equipment that goes into space, public discussion is stifled. The time for national debate about the direction of our space policy is now, not after we are inexorably committed to a new arms race in space.

It is in the hope of opening that debate, not of settling it, that I offer the following theses.

SPACE LASER WEAPONS WILL NOT PROTECT US FROM THE THREAT OF NUCLEAR WAR.

Neither side is going to give up nuclear 'deterrence.' Each will find the countermeasures or build the new weapons necessary to maintain its ability to destroy the other. Anti-ballistic missile systems in space or on the ground will stimulate new rounds in the nuclear arms race, not end it.

On the other hand, space laser weapons offer an excellent opportunity for preventive arms control. Both sides have better uses for the hundreds of billions that these weapons would cost. And verifying compliance with an agreement not to deploy the massive space weapons platforms would be extremely easy.

WITH OR WITHOUT NEW SPACE SYSTEMS, WE CAN'T WIN A NUCLEAR WAR.

Military men are properly concerned that our side should retain the ability to retaliate against a Soviet nuclear attack, and even that some choices are available as to exactly what kinds of targets to retaliate against. But retaliation is one thing and 'nuclear war fighting' is another. Our Secretary of Defense has said that we don't want the ability to 'win' a nuclear war, just the ability to 'prevail' – a distinction he has failed to clarify. No matter how much we spend on 'strategic connectivity' and 'satellite survivability,' a nuclear war will be the end of the game. It's dangerous to think otherwise.

WE CAN'T HOPE TO MAKE THE EARTH SAFE FROM WARFARE BY MOVING COMBAT INTO SPACE.

Military systems in space are designed to produce military advantages on the ground. Navies have not made land armies obsolete. Air forces have not made navies or armies obsolete. New forms of combat have complemented the old forms, not replaced them.

If we and the Soviet Union could agree to limit our battles to the space arena, we could just as well agree to limit them to a deserted Pacific island. We could send representative knights to joust for us, then agree to abide by the outcome. The reasons we don't do that are the same reasons we won't have technological duels isolated in space.

ONLY IN A LIMITED SENSE IS SPACE THE NEW 'HIGH GROUND.'

In the sense that you can see more of space and send messages farther across it, it is, but in the sense that you could lob weapons down from space while the enemy has an upward struggle, it isn't. Neither we nor the Soviets are likely to be able to 'seize and hold' space as though it were some strategic territory.

Space systems do offer 'force multiplying' benefits to our land, sea, and air forces. These benefits will be most useful against smaller, weaker countries far from our shores. Against the other major space power, the Soviet Union, the benefits will be offset somewhat by the fact that the Soviets have access to the same kinds of technology.

WE SHOULD THINK TWICE BEFORE TRYING TO 'SHAPE THE
MILITARY COMPETITION' BY MOVING WEAPONS INTO SPACE.

Many military men agree that the United States is more
dependent on its space system than is the Soviet Union.
But some military men talk of 'control' and 'denial.' The
goal is to be able to protect our own military use of space
while denying its use to the other side. But it's by no means
clear that this goal is feasible. For the foreseeable future,
'denial' – the destruction of satellites – is going to be
easier than 'control.' And we may have more to lose.

An unfettered military competition in space could leave
the U.S. worse off, whether its technology was superior or
not. For their part, the Soviets have shown a willingness to
make whatever sacrifices are necessary to be seen as the
military equals of the United States.

At great cost, we might win some temporary advantages
by accelerating the military space technology race. But as
sick as the Soviet economy may be, ours is hardly bloom-
ing. The idea that we can apply our own scarce resources so
as to force the Soviet Union into either military inferiority
or economic collapse doesn't wash.

WE WILL PAY A PRICE FOR THE INCREASING MILITARIZATION OF
OUR NATIONAL SPACE PROGRAM.

When President Eisenhower and Congress created the
National Aeronautics and Space Administration, they
established a policy of giving predominance to the civil
exploration and exploitation of space. We have reaped
prestige, national pride, economic benefit, and technological 'spin-offs' from that policy.

Now, the military space budget is $2 to $4 billion a year
larger than NASA's. Our civilian space activities derive
few benefits from military space developments (the
amount of 'technology transfer' is low, according to the
Congressional Office of Technology Assessment). The
General Accounting Office has pointed out that a fairer
accounting would identify about 25 percent of the NASA
budget as supporting military purposes as well.

As a result of this shift in resources, not only our space exploration programs but also research and development in communications and earth resources monitoring satellites are suffering as well. The Office of Technology Assessment has pointed out that some of our major economic competitors – notably the French and the Japanese – are moving in to exploit technology we have developed for their own commercial benefit. As we struggle to pull ahead in the military race in space, we are in danger of falling behind in the civil one.

# NOTES

1. Pages 2–3, 'There was an electricity . . .': De Witt S. Copp, *A Few Great Captains: The Men and Events That Shaped the Development of U.S. Air Power* (Garden City, N.Y.: Doubleday & Company, 1980), p. 37.
2. Page 3, Mitchell '. . . spoke on the unmentionable . . .': *Captains*, p. 38.
3. Page 9, Footnote, O'Malley, 'Remarks to the Manned Space Flight Support Group,' Johnson Space Center, Texas, November 1, 1980. USAF News Release.
4. Page 10, It would . . . : purposes of the Space Directorate described in 'New USAF Organization to Intensify Space Focus,' *Aviation Week and Space Technology*, October 26, 1981, p. 25.
5. Page 10, 'I believe the right answer . . .': Edward Aldridge, Address to the National Space Club, November 18, 1981, speech text.
6. Page 16, '. . . in the not too distant future . . .': Lieutenant Colonel Dino A. Lorenzini, USAF, and Major Charles L. Fox, USAF, '2001: A U.S. Space Force,' *Naval War College Review*, March/April 1981, p. 64.
7. Page 19, 'My loss rate . . .': 'View from the Top: Lieutenant General Richard C. Henry,' interview in *Military Electronics and Countermeasures*, July 1981, p. 22.
8. Page 21, 'What we have not been able to do . . .': Henry, 'View from the Top,' same interview, p. 21.
9. Page 21, In November 1971 . . . : the narrative following relies heavily on a company history, *The Aerospace Corporation: Its Work: 1960–1980* (El Segundo, Cal.: The Aerospace Corporation, 1980), pp. 56 and following.
10. Page 26, 'Today, spacecraft . . .': Henry, 'View from the Top,' cited above, p. 21.
11. Pages 36–7, 'The space shuttle is to a significant degree . . .': Carl Sagan, *The MacNeil-Lehrer Report*, Public Broadcasting System, November 4, 1981.
12. Page 45, Brigadier General Spence Armstrong, USAF, 'Presentation to the Committee on Armed Services,' House of Representatives, April 1980 (Mimeo).
13. Page 62, Dineen pointed out . . . : Gerald Dineen, testimony in U.S. Congress, House, *Military posture and H.R. 10929*, Hearings before the Committee on Armed Services, House of Representa-

tives, 95th Cong., 2nd sess., Part 3, Book 2, *Research and Development, Title II*, p. 1844.

**14.** Page 74, '. . . I want to state at the outset . . .': Vice Admiral Gordon Nagler, testimony in U.S. Congress, House, *Navy Leased Satellite (LEASAT) and Fleet Satellite (FLTSAT) Programs*, Hearing before the Committee on Armed Services, House of Representatives, 97th Cong., 1st sess., June 23, 1981, p. 5.

**15.** Pages 74–5, 'A return to non-satellite communications . . .': Nagler, same place, p. 9.

**16.** Page 75, 'While the potential missions . . .': Report of Secretary of Defense Harold Brown to Congress, January 19, 1981, reprinted in U.S. Congress, Senate, *Department of Defense Authorization for Appropriations for Fiscal Year 1982*, Hearings before the Committee on Armed Services, United States Senate, 97th Cong., 1st sess., p. 167.

**17.** Pages 81–2, '. . . we have come a long way . . .': General Richard Ellis, USAF, testimony in U.S. Congress, Senate, *Department of Defense Authorization for Appropriations for Fiscal Year 1982*, Hearings before the Committee on Armed Services, United States Senate, 97th Cong., 1st sess., *Part 7, Strategic and Theater Nuclear Forces, Civil Defense*, p. 4212.

**18.** Page 82, 'Once execution is complete . . .': Ellis, same place, pp. 4215–4216.

**19.** Pages 83–4, 'The evolution of doctrine . . .': John Morgenstern in *Signal*, December 1980, p. 57.

**20.** Page 84, '. . . I remain highly skeptical . . .': Brown, Report of Secretary of Defense Harold Brown to Congress, cited above, p. 127.

**21.** Pages 84–5, 'In previous years the concept . . .': Stansberry quoted by Clarence A. Robinson, Jr., 'Pentagon Backs Strategic Modernization,' *Aviation Week and Space Technology*, October 26, 1981, p. 53.

**22.** Page 86, '. . . this [FLTSATCOM] is only . . .': Randerson testimony in U.S. Congress, House, *Navy Leased Satellite . . .*, cited above, p. 18.

**23.** Page 88, 'I think the judgment . . .': Hans Mark, testimony in U.S. Congress, House, *Department of Defense Appropriations for 1981*, Hearings before a subcommittee of the Committee on Appropriations, House of Representatives, 96th Cong., 2nd sess., *Part 2*, p. 356.

**24.** Page 89, 'Milstar is designed . . .': Major General Gerald K. Hendricks, 'Electronics and Military Space Systems,' Presentation for Air Force Association Electronics in the Air Force Conference, April 26, 1982, Wakefield, Mass. (photocopy), p. 7.

**25.** Pages 104–5, 'major sensor camera . . .': Harry V. Martin, 'Electronics Remains Keystone to U.S. Intelligence Mission,' *Defense Electronics*, December 1981, p. 74.

**26.** Page 107, 'there are 134 major final assembly plants . . .': statement of DIA officials Major General Richard X. Larkin and Edward M. Collins in U.S. Congress, House and Senate, *Allocation of Resources in the Soviet Union and China – 1981*, Hearings before a subcommittee of the Joint Economic Committee, 97th Cong., 1st sess., p. 83.

27. Page 111, Footnote, The NRO operates . . .: this information from Harry F. Eustace, 'Changing Intelligence Priorities,' *Electronic Warfare/Defense Electronics*, November 1978, pp. 40 and 82.

28. Page 117, . . . if one were to print a map . . .: *Aviation Week and Space Technology*, January 11, 1981, p. 51.

29. Page 119, 'capable of performing 40 million . . .': *Aviation Week and Space Technology*, February 16, 1981, p. 59.

30. Page 119, 'This processor was initially designed . . .': testimony of Robert Fossum in U.S. Congress, House, *Military Posture and H.R. 2970*, Hearings before the Committee on Armed Services, House of Representatives, 97th Cong., 1st sess., *Part 4, Research and Development, Title II*, p. 154.

31. Page 125, 'It was at night . . .': William Perry, testimony in U.S. Congress, Senate, *Department of Defense Authorization for Appropriations for Fiscal Year 1981*, Hearings before a subcommittee of the Committee on Armed Services, United States Senate, 96th Cong., 2nd sess., *Part 5, Research and Development*, p. 2674.

32. Pages 127–8, In 1958 the Navy hired . . .: history of the Transit system found in Geoff Richards, 'Transit–The First Navigational Satellite System,' *Spaceflight*, February 1979, pp. 50–55.

33. Page 135, '. . . transpond through a device in the missile . . .': Rear Admiral Robert Wertheim, testimony in U.S. Congress, House, *Department of Defense Appropriations for 1982*, Hearings before a subcommittee of the Committee on Appropriations, House of Representatives, 96th Cong., 2nd sess., *Part 3*, p. 387.

34. Page 136, 'What a tremendous asset . . .': Richard Henry, 'View from the Top,' cited above, p. 14.

35. Page 139, 'Some people argue . . .': William Perry, quoted in *Aerospace Daily*, June 12, 1980, p. 234.

36. Page 140, 'One should be able to predict . . .': Eberhardt Rechtin, 'What Next? The Military Applications Outlook,' Address to the American Institute of Aeronautics and Astronautics, May 14, 1981 (The Aerospace Corporation print), p. 5.

37. Page 140, 'the pace of battle . . .: Lieutenant General Donald Keith, testimony in U.S. Congress, Senate, *Department of Defense Authorization for Appropriations for Fiscal Year 1982*, Hearings before the Committee on Armed Services, United States Senate, 97th Cong., 1st sess., pp. 1385–1386.

38. Page 152, The new Air Force ASAT . . .: Calculations on the likely velocity of the U.S. ASAT by John Pike, 'Performance of the New American Air-Launched ASAT,' Federation of American Scientists Staff Research Note, 5 July 1983.

39. Page 158, 'The ability of the Soviet Union . . .': Brigadier General

Ralph Jacobson, testimony in U.S. Congress, House, *Department of Defense Appropriations for 1981*, Hearings before a subcommittee of the Committee on Appropriations, House of Representatives, 96th Cong., 1st sess., *Part 1*, p. 880.

40. Page 158, '. . . important not to couple . . .': Seymour Zeiberg testimony quoted in Craig Couvault, 'Anti-satellite Weapon Design Advances,' *Aviation Week and Space Technology*, June 16, 1980, p. 247.

41. Page 161, 'It is not possible at present . . .': William Perry, testimony in U.S. Congress, House, *Department of Defense Appropriations for 1981*, Hearings before a subcommittee of the Committee on Appropriations, House of Representatives, 96th Cong., 2nd sess., *Part 3*, p. 501.

42. Page 161, 'I don't think we have the right ideas . . .': Robert Cooper quoted in *Aviation Week and Space Technology*, February 1, 1982, p. 23.

43. Page 163, '. . . equipped with firing ports . . .': *Aviation Week and Space Technology*, November 30, 1981, p. 17.

44. Page 163, The *Aviation Week* editors later . . .: *Aviation Week and Space Technology*, March 8, 1982, p. 33.

45. Page 165, '. . . it's a possibility, but . . .': AP report in *Philadelphia Inquirer*, May 23, 1980, p. 3.

46. Page 165, 'The USSR already has . . .': *Aviation Week and Space Technology*, June 16, 1980, p. 60.

47. Page 166, 'laser hardening and sensor protection . . .': Colonel Donald Henderson, 'Defending Our Space Assets – the Issues and the Challenges,' paper prepared for a convention of the American Astronautical Society, San Diego, October 1981 (undated photocopy).

48. Page 182, Some argue that one laser battle station . . .: details here from a speech by Senator Malcolm Wallop, *Congressional Record*, Senate, July 1, 1980, pp. S-9074–9079.

49. Page 184, 'an unknown that has to be well understood': Richard Carlson, testimony in U.S. Congress, Senate, *Laser Technology – Development and Applications*, Hearings before the Subcommittee on Science, Technology, and Space of the Committee on Commerce, Science and Transportation, U.S. Senate, 96th Cong., 1st and 2nd sess., 1979 and 1980, p. 192.

50. Page 185, 'Making such a mirror . . .': Kosta Tsipis, 'Laser Weapons,' *Scientific American*, December 1981, p. 55.

51. Page 185, 'there has never been any doubt . . .': Wallop quoted in *Aviation Week and Space Technology*, May 25, 1981, p. 53.

52. Page 185, 'to resolve any remaining basic . . .': 'Laser Battle Station Mirror Proposed,' *Aviation Week and Space Technology*, May 25, 1981, p. 64.

53. Page 186, 'Advanced space-based surveillance . . .': Dr. James P. Wade, Assistant to the Secretary of Defense for Atomic Energy and Acting Principal Deputy Under Secretary of Defense for Research

and Engineering (Strategic and Space Systems), testimony in U.S. Congress, Senate, *Department of Defense Authorization for Appropriations for Fiscal Year 1982*, Hearings before the Committee on Armed Services. U.S. Senate, 97th Cong., 1st sess., 1981, *Part 7*, p. 4115.

54. Pages 187–8, 'Another risk is operational . . .': George Millburn, testimony in Senate Armed Services Committee hearings above, p. 4145.

55. Page 188, '. . . if we considered the Soviet numbers . . .': Millburn, same testimony, p. 4152.

56. Pages 188–9, 'There is no question but what the compelling . . .': Verne Lynn, same place, p. 4159.

57. Page 189, 'The survivability of the laser . . .': Millburn, same place, p. 4165.

58. Page 195, 'technically, we are really not talking . . .': this exchange appears in the same place on p. 4155.

59. Page 196, 'not regard itself to be in a race . . .': Richard Airey in U.S. Congress, Senate, *Laser Technology – Development and Applications*, Hearings before the Subcommittee on Science, Technology, and Space of the Committee on Commerce, Science, and Transportation, U.S. Senate, 96th Cong., 1st and 2nd sess., 1979 and 1980, p. 86.

60. Page 197, 'Early technology demonstration . . .': Clarence Robinson, 'Laser Technology Demonstration Proposed,' *Aviation Week and Space Technology*, February 16, 1981, p. 16.

61. Pages 199–200, 'In the 21st Century directed energy weapons . . .': Warner memo in *Congressional Record – Senate*, May 13, 1981, p. S4977.

# ACKNOWLEDGMENTS

Unless otherwise indicated by endnotes, the quotations in this book come from personal interviews. My thanks to those Air Force officers, Defense Department officials, and industry officials who generously shared their time and ideas with me. I promised them that although my conclusions might differ from theirs, I would try to give their view a fair airing. I hope they are not disappointed.

I also promised to emphasize, and I do so here – that in most cases the views quoted reflected the personal opinions of those interviewed, not the official policies of the organizations employing them.

Credit is due as well to those public affairs personnel at Air Force headquarters, the Office of the Secretary of Defense, the Air Force Space Division, the Aerospace Defense Command, and the Air Force Academy who helped arrange visits and interviews for me and who provided most of the illustrations in the book.

For helping to shape the proposal that became a book, I thank Raphael Sagalyn. For reading and commenting on parts of the manuscript, thanks to Lucille Schultz and Kerry Joels.

For giving me an institutional home during my lonely sojourn as a 'free-lancer,' thanks to Jeremy Stone and the Federation of American Scientists.

Thanks to my editor, Alice Mayhew, and her hardworking assistant editor, David Masello, for helping to make this book the one I wanted to write. Patricia Miller's copy editing was admirably competent. Maria Iano did an enterprising job of putting together the picture section. Responsibility for errors of fact or omission, of course, remain my own.

# INDEX

# JUDITH COOK

## RED ALERT

Nuclear power: 'The safest form of energy known to man'

*Peter Walker, Energy Minister*

That's the official line. And for thirty years most people swallowed it. Doubters were cranks or scaremongers.

But then the reports began to build up. It slowly became clear that Windscale (hastily renamed Sellafield) had a long history of accidents, fires, radioactive leaks. Anxiety grew.

**Then came Three Mile Island**
The Sizewell 'B' Enquiry was set up. Nirex began to roam the countryside looking for nuclear dumping sites.

**Then came Chernobyl**
Now at last in *Red Alert* is the *true* history of the nuclear power industry worldwide, the incredible record of accident and cover-up, mismanagement and political manipulation, of facts suppressed and downright lies.

POST A LITTLE HAPPINESS

# Post·A·Book

A Royal Mail service in association with the Book Marketing Council & The Booksellers Association.

Post-A-Book is a Post Office trademark.

# JAN MOEN

## JOHN MOE DOUBLE AGENT

'I handed him my pistol, and said as calmly as I could, "We were put ashore as German spies."'

That surrender to a startled Scottish policeman in 1941 marked the start of one of the longest lasting and most successful deceptions of World War II.

Trained by the Germans in espionage and sabotage techniques, John Moe was to spend much of the rest of the war on behalf of M15, transmitting disinformation back to his unsuspecting German 'controllers'.

He travelled the length and breadth of Britain. He reported on troop movements and military equipment. He even seemingly carried out acts of sabotage. His reports played a major part in convincing the Germans that Norway rather than Normandy would be the objective of the Allied invasion of 1944 – so keeping thousands of enemy troops tied down uselessly in the wrong place.

Now living quietly in Sweden, John Moe has told for the first time the extraordinary story of his wartime life as a double agent.

**NEW ENGLISH LIBRARY**

# TAM DALYELL

## MISRULE

*Mrs Thatcher has lied.*

*The Belgrano sinking, the Westland affair, the US air raid on Libya, the miners' strike, the Zircon TV programme and the police raid on BBC Glasgow, Peter Wright and the Spycatcher business ...*

Again and again, quite deliberately, Mrs Thatcher has lied.

Which is why, day-in, day-out, year-in, year-out, Tam Dalyell comes back again and again to the same points, picking away, questioning ...

The Government, and Our Leader in particular, would dearly like us to believe that the man's just some boring obsessive who doesn't know when to shut up. But in *Misrule*, his central charge is made clear: we have a Prime Minister who lies and who orders lies to be told, not for weighty reasons of state, but simply to save her own skin.

*That is Tam Dalyell's charge and this is the evidence.*

'Remarkable and important'
> *The Times Literary Supplement*

'Essential ammunition'

> *The Independent*

**NEW ENGLISH LIBRARY**

## MORE NON-FICTION FROM
## HODDER AND STOUGHTON PAPERBACKS

### JAN MOEN
☐ 41909 6  John Moe Double Agent                £3.50

### NIGEL WEST
☐ 41984 9  Molehunt                             £3.50
☐ 33781 8  A Matter of Trust                    £2.95
☐ 41197 X  GCHQ                                 £3.95

### JUDITH COOK
☐ 41379 9  Red Alert                            £3.50
☐ 05885 9  Who Killed Hilda Murrell?            £1.95

### TAM DALYELL
☐ 42488 X  Misrule                              £2.95


*All these book are available at your local bookshop or newsagent, or can be ordered direct from the publisher. Just tick the titles you want and fill in the form below.*

Prices and availability subject to change without notice.

---

Hodder & Stoughton Paperbacks, P.O. Box 11, Falmouth, Cornwall.

Please send cheque or postal order, and allow the following for postage and packing:

U.K. – 55p for one book, plus 22p for the second book, and 14p for each additional book ordered up to a £1.75 maximum.

B.F.P.O. and EIRE – 55p for the first book plus 22p for the second book, and 14p per copy for the next 7 books, 8p per book thereafter.

OTHER OVERSEAS CUSTOMERS – £1.00 for the first book, plus 25p per copy for each additional book.

NAME.................................................................................

ADDRESS .........................................................................

.........................................................................................